The Art of Science

Rossella Lupacchini • Annarita Angelini
Editors

The Art of Science

From Perspective Drawing to Quantum Randomness

 Springer

Editors
Rossella Lupacchini
Annarita Angelini
University of Bologna
Bologna, Italy

ISBN 978-3-319-37907-4   ISBN 978-3-319-02111-9 (eBook)
DOI 10.1007/978-3-319-02111-9
Springer Cham Heidelberg New York Dordrecht London

Printed on acid-free paper

Springer is part of Springer Science+Business Media (www.springer.com)

Every truth requires some pretence to make it live.

Joseph Conrad

Preface

The Renaissance is famous for its discovery of linear perspective, complex numbers, and probability. History has been quick to recognize the power of perspective that gave form to a "classic" style in painting, but has failed to acknowledge the true significance of complex numbers and probability. Both were treated with a great deal of suspicion by the scientific establishment and as a result were overlooked for many years. Linear perspective was already four centuries old, when quantum theory first showed how probability might be moulded from complex numbers and went on to create the notion of "complex probability amplitude". Yet, from a theoretical point of view, the space opened by linear perspective to painting and the space opened by complex numbers to science are equally important and share many characteristics. This book explores that shared field.

It may well seem challenging, or even inappropriate, to relate notions belonging to contemporary science with inventions and themes of the Renaissance. But we want to make it clear that we have no wish to antedate the findings of the contemporary science back into the Renaissance, nor to trace the history of science from the fifteenth century to the present. Instead, our purpose is to extend the "ideal" style Leonardo conceived for painting to science. In Leonardo's view, painting must recreate the geometry of nature through the harmony of form; "the mind of the painter must transmute itself into the very mind of nature and be the interpreter between it and art" (*Trattato*, I, 24v). Our ambition is to encourage the reader to see science as an "art" and art as a form of scientific knowledge.

As to the title, *The Art of Science* reverses *The Science of Art* by Martin Kemp as it envisages a "complementary" view. *The Science of Art* (1992) rests on the premise that there *were* special kinds of affinity between art and science from the Renaissance to the nineteenth century, and on the observation that "the affinities centred upon a belief that the direct study of nature through the faculty of vision was essential if the rules underlying the structure of the world were to be understood." Consequently, Kemp's book focuses on *optically minded* theory and practice of art. Its primary concern is to examine the extent to which artists' work and ideas were scientifically founded. *The Art of Science*, instead, rests on the premise that there *are* special kinds of affinity between art and science and sees the affinities emerging from a conception of art and science as "symbolic forms". Accordingly, the faculty of vision is essential as it turns "imagination" into visual and graspable forms. In Leonardo's words: "[The eye] triumphs over nature, in that the constituent parts of nature are finite, but the works which the eye commands of the hands are infinite, as is demonstrated by the painter in his rendering of numberless forms." To master the rules of Albertian perspective allows the painter not only to depict the world as it appears but also to *see and draw other possible worlds*. This is the main lesson that science gains from the Renaissance art.

The philosophical concerns underlying our project are sympathetic to attempts to revise the "picture theory of science" (*Bildtheorie*) and, in a broad sense, to a "structuralist" view of science. While we will not enter into the contemporary debate about the themes of structuralism, we want to pay tribute in retrospect to two leading figures: Ernst Cassirer and Hermann Weyl.

The heritage of Cassirer's *Philosophy of Symbolic Forms* (1923–1929) cannot be confined within the main stream of the neo-Kantian philosophy, *tout court*. His revision of the "transcendental" approach highlighted a common denominator among a variety of "forms" arising in remote disciplines and cultural areas. This shared term, which manifests a "symbolic" character, allows Cassirer to compare the extraordinary variety of products of human spirit (myth, language, art, science) and to understand all of them as symbolic constructions in the general frame of a "science of culture" (*Kulturwissenschaft*): "The fundamental concepts of each science, the instruments with which it propounds its questions and formulates its solutions, are regarded no longer as passive images of something given but as *symbols* created by the intellect itself." The search for a theory of artwork within a comprehensive *Kulturwissenschaft* and the attempt to deduce the meaning of symbols created by art from their iconographic content and style may bring to mind the iconological researches developed by Aby Warburg and his Circle (joined, among the others, by Fritz Saxl and Ernst H. Gombrich). The Warburg programme, however, was more historical than theoretical. Even when an interpretation was advanced—such as Erwin Panofsky's *Perspective as Symbolic Form* (1924–1925)— the paradigm was borrowed from a theory of knowledge external to the artistic work and style under consideration. Hence, the artistic representation was to provide evidence for a previously accepted theory. By contrast, our goal is to focus on the artistic "invention" at the beginning, not at the end, of a theoretical path that leads to the scientific representation. In this way, through the medium of mathematical thought, a "visual" form can be used as a model for scientific knowledge.

As Leonardo's pictorial style is related to the geometry of nature, so is Hilbert's mathematical "style" related to his vision of geometric forms. The "ideal" style Hilbert conceived for mathematical knowledge results in a "general theory of forms". In particular, if we look for evidence of our claim that "the faculty of vision is essential as it turns 'imagination' into visual and graspable forms", we should pay attention to his essays on "intuitive geometry". To fully appreciate the potentialities embedded in Hilbert's picture of mathematical theories and their impact on the development of physical concepts, we should look at Hermann Weyl's writings. While his refined works on mathematical physics—such as *The Theory of Groups and Quantum Mechanics* (1931)—disclose a "visual understanding" of science only to scientists, his *Philosophy of Mathematics and Natural Science* (1927–1949) enhances mutual understanding between humanities and science as it shows the symbolic form of their specific contents. Finally, his *Symmetry* (1952) is a model to follow for an "art guide" to science.

This book has a long story. A shared interest in conceptual and epistemological issues relevant to art and science prompted us to conceive a project on *Reality and Its Double. Perspective and Complex Numbers Between the Renaissance and Quantum Physics*, awarded by the *Istituto di Studi Avanzati* (ISA) of the University of Bologna in 2009. During the events connected with the project including a series of lectures and a closing conference on *The Art of Science*, we had the opportunity to discuss and clarify issues and consequently select the most relevant topics. This volume includes papers delivered both as lectures and as contributions to the conference plus some that were specially commissioned.

We are grateful to the *Istituto di Studi Avanzati* for its generous support of our project and to all the participants for their valuable contributions. In particular, we want to thank John Stillwell for his unwavering confidence in the idea of this book. We are also grateful to the reviewers for their comments on the manuscript. Finally, it is a pleasure to thank Giuseppe Longo and Wilfried Sieg for their gentle encouragement and David Deutsch for his valuable comments and suggestions.

Bologna, Italy
September 2013

Rossella Lupacchini
Annarita Angelini

Contents

Part I
Ways of Perspective

What does artificial perspective tell us about scientific knowledge? How does it enhance "visual understanding" of mathematic forms? How does it refine the notion of "observability"? By addressing such questions from different points of view—mathematical, philosophical, historical—the essays collected in Part I lead to the depiction of linear perspective as an art of seeing, projecting, and measuring.

In the opening chapter, John Stillwell directs us to view *perspectiva pingendi* with a mathematician's eye. The discovery of the *costruzione legittima* for perspective drawing—namely, of a method for projecting the three-dimensional space on the pictorial surface and, more in general, of a "scientific", optical system allowing any *imaginary* scene to be represented as if it were *real*—led to interest in a new kind of geometry, *projective geometry*, in which points and lines are the main ingredients. Thanks to the possibility of creating perspective drawings without measurement, projective geometry freed itself completely from coordinates and became a system in which all theorems were derived by reasoning about points and lines alone. Seemingly, geometry and algebra had diverged completely. But a surprising development was on the horizon: when geometry is freed from numbers, addition, and multiplication, it becomes feasible to reconstruct algebra on a purely geometric foundation by means of the Pappus theorem, the Desargues theorem, and the little Desargues theorem. Even more surprisingly, these purely geometric theorems were found (by David Hilbert and Ruth Moufang) to control which kind of algebra is possible in two, four, and eight dimensions.

If, on the one hand, linear perspective encouraged mathematicians to see a new kind of geometry, on the other, the medieval interpretations of the Euclidean geometric optics encouraged Renaissance painters to play with its rules. After flying towards the eighth dimension, Nader El-Bizri takes us back to see "reality" in perspective. His essay (Chap. 2) contrasts dialectically the "art" of optics with the "science" of painting. The pictorial structure is intrinsically implied within the visual elements of the science of optics and geometry, while it simultaneously depends on these sciences for the projections and constructions needed to render spatial depth in artificial perspective. The "science of painting" is set against the principal classical theories of optics and geometry. Even though Renaissance authors were more often theoretically inclined to follow Euclid, Ptolemy, and Vitruvius, they nonetheless paid much attention to the transmitted traditions that were associated with the eleventh century Arab polymath al-Hasan Ibn al-Haytham (known as Alhazen). Adaptively mediated by medieval European opticians and mathematicians, they led ultimately to the transformation of "natural" perspective into "artificial" perspective; hence, to teach painting how to *imitate* "reality" bypassing the natural vision.

It was Filippo Brunelleschi's work, particularly in architecture, that inaugurated a most radical deviation from late medieval tradition. It endorsed, in the most striking way, the widespread humanistic intolerance of Scholastic scientific conception (Chap. 3). The internal organization of Brunelleschi's buildings showed entirely new optical unity of space, precisely defined architectural elements, emphasis on the visible manifestation of proportions and, what is most radical, lack (negation) of paintings and colours on walls. The white surface works as a "transcendental" light, providing a background for the primary elements of the buildings (columns, arches, architraves) made of darker stone (*pietra serena*). Dalibor Vesely explores the meaning of Brunelleschi's primary architectural elements, their relation to Alberti's *lineamentum*, and also to Zuccaro's and Mannerists' *disegno interno*. These relations appeared to be supported by a neo-Platonic metaphysics of light and its epistemological consequences. As a "universal formative power", *disegno interno* may be viewed as a general source of creativity underlying modern forms of knowledge and, consequently, modern European culture as a whole.

From a mathematical point of view, the method of 'ideal elements' demonstrates that universal formative power. Indeed, Hilbert traced the origin of the ideal elements to the points at infinity of plane geometry. Such ideal points where parallel lines meet on the projective plane originate as vanishing points on the pictorial plane. Before the independence of the parallel postulate was 'logically' proven, an ideal non-Euclidean geometry was 'visually' presented by perspective drawing. Therefore, taking art's imagination to its limits, mathematics has produced new forms of 'visual' geometry. Tristan Needham (Chap. 4) not only drives us to see visual differential geometry as an artwork, but also depicts its forms with a painter's hand. Beltrami's interpretation of the hyperbolic geometry comes to life with rare intensity in the figures accompanying the text.

Although artistic and scientific knowledge may seem to go hand in hand in the Renaissance, their relationship may appear controversial as much in modern and contemporary culture as in ancient thought. In Victor Stoichita's *Short History of Shadow* (1997), both the myth regarding the birth of artistic representation, in Pliny's *Natural History*, and the myth regarding the birth of cognitive representation, in Plato's cave, are traced to the motif of shadow. According to Pliny the Elder, painting originated from the idea of circumscribing shadows by lines (*omnes umbra hominis lineis circumducta*). It was a young woman in love who, when her lover was going abroad, "drew in outline on a wall the shadow of his face thrown by the lamp" (*Natural History*, XXXV, 35,151). For Plato, however, a shadow has a "negative" connotation turning "what is observable" into a dark spot, a "phantom" (*eídolon*). Seeing nothing but projected shadows, the prisoners in the cave took shadows for reality. Their "cognitive" representation may be compared with that of the painter whose art is directed to the imitation of appearances (*phantasma*) not of truths (*aletheia*): "the mimetic art is far removed from truth," observes Socrates in *The Republic*, "and this, it seems, is the reason why it can produce everything, because it touches or lays hold of only a small part of the object, and that a phantom" (*Rep.* 598b). Even the Eleatic Stranger, reporting Plato's thought in *The Sophist* (236c-e), distinguished a "fantastic art"—the art of producing appearances and presenting them as if they were real things (*tékne phantastiké*), i.e., painters' and Sophist's art—from a *less imperfect* "likeness-making art", the art of producing copies (*tékne eikastiké*). Thus painting was confined to the bottom of Plato's cave, while "science", as an imitation of the truth, aimed at producing copies of reality as it is.

Since Plato, "scientific knowledge" has not been concerned with the description of "shadows", and even less with the production of fantastic images. Its principles cannot be reconciled with the essentially plural and "sophistic" character of painting. Its images must convey a "veritable" and "realistic" *mimesis*. Indeed, it ascribes to them the same quality of "specular-reflection" that Socratic sapience ascribed to the self-knowledge of soul. A scientific representation is conceived as a mirror of reality and distinguished from deceptive appearances. Accordingly, a scientist plays the role of a neutral observer and assesses the degree of similarity between the "truth" of the observed reality and the "truth" of the scientific representation. Yet, a mirror image immediately evokes Narcissus' metamorphosis which, on reflection, is a consequence of a deception. Narcissus falls in love with his own specular-image, believing it to be the "shadow" of someone else. "That which you behold is but the shadow of a reflected form (*ista repercussae, quam cernis, imaginis umbra est*)" (*Metamorphoses*, 3, 436). The seduction of the *other* becomes a first step towards the recognition of one's own *self* reflected in the mirror. "I burn with love of my own self; I both kindle the flames and suffer them" (*ibid.*, III, 464). Though scientific knowledge aims at mirroring reality, in its historical development, the awareness of the "action" of mirrors has been oscillating: from the maximum of illusion, according to which scientific representation is the faithful image of reality reflected in a mirror, to the maximum of narcissistic disenchantment (or enchantment), according to which scientific representation mirrors scientist's "vision", unveiling the logic underlying the construction of knowledge.

In the Albertian perspective, the artistic representation born from shadow is engaged with the "mirror-reflection" pursued through the scientific tradition. The art of painting then performs a dual magic: as a *shadow*, a simulacrum of a lack of sensibility, it is more "powerful" than a direct (sensory) vision depending on the body's "measurability" constraints; as a *reflection* of a point of view, namely, of a subjective criterion (*ratio*), it relates the "resemblance" between the original and the copy to the artist (Fig. 1). As Narcissus is at the same time subject and object of his desire, so the artist, or that "layman of wisdom" which dominates the Renaissance

Fig. 1 Giorgio Vasari: *The Studio of the Artist, c.* 1563. Florence, Casa Vasari

scene, is aware that "the object is now something other than the mere opposite the—so to speak—*ob-jectum* of the Ego. It is that towards which all the productive, all the genuinely creative forces of the Ego are directed".[1]

Simon Altmann (Chap. 5) sheds light on the action of mirrors in art and science. Indeed, since Narcissus was seduced by his mirror image and turned into a flower, humanity has been both fascinated and concerned by mirrors. Although art proceeded rather slowly from mirror symmetries (notable examples can be seen Greek pottery decorations) to more complex rotational patterns, it was not until the nineteenth century that the mathematics of rotations was understood. From specular and rotational patterns emerged the mathematics of *quaternions* and *spinors*, which eventually, influenced profoundly our knowledge of physics, especially quantum physics.

---

[1]E. Cassirer, *Individuum und Kosmos in der Philosophie der Renaissance* (1927). English translation: *The Individual and the Cosmos in Renaissance Philosophy*, Univ. of Pennsylvania Press, Philadelphia 1963, p. 143.

Chapter 1
From Perspective Drawing to the Eighth Dimension

John Stillwell

1.1 Problems of Perspective

The *Arnolfini Portrait* (Fig. 1.1), by Jan van Eyck (1434), is an acclaimed example of the new realism in Flemish painting in the early fifteenth century. It seems to be an accurate depiction of a three-dimensional space and of three-dimensional objects. However, van Eyck's treatment of perspective is not mathematically correct.

Take a closer look at the chandelier (Fig. 1.2).

If the six arms of the chandelier are identical, then the lines connecting corresponding points on the two arms must be parallel. We consider such points on the two leftmost arms. Figure 1.3 shows the lines connecting three pairs of corresponding points—one line through the tops of the candle holders, and lines through the first and second crockets.

Parallels can't look like this! They should either look parallel or else converge to a common point "at infinity."

There are fifteenth century artworks with far more blatant errors in perspective than Jan van Eyck's. Figure 1.4 shows one, from the unknown illustrator of a book by Savonarola (ca. 1497).

By trying to make parallels "look parallel" when they should meet at infinity, the artist has lost control of another set of parallels, which are not even straight! This error brings to light a key problem in perspective—drawing a tiled floor.

J. Stillwell (✉)
University of San Francisco
e-mail: stillwell@usfca.edu

R. Lupacchini and A. Angelini (eds.), *The Art of Science*,
DOI 10.1007/978-3-319-02111-9__1,
© Springer International Publishing Switzerland 2014

Fig. 1.1 Jan van Eyck: *The Arnolfini Portrait*, 1434. London, National Gallery

Fig. 1.2 The chandelier in the *Arnolfini Portrait*

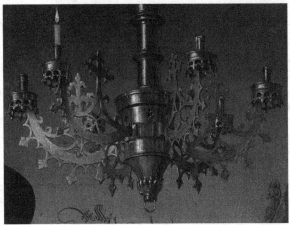

1.1.1 Drawing a Tiled Floor in Perspective

One of the first artists to understand the mathematics involved in perspective drawing was Piero della Francesca, whose *Flagellation of Christ*, from around 1460 (Fig. 1.5), includes a meticulously drawn tiled floor.

Fig. 1.3 *Lines* through corresponding points on two arms

Fig. 1.4 Bartolomeo di Giovanni: Illustration from Savonarola's *Dell'Arte di Ben Morire*

Fig. 1.5 Piero della Francesca: *Flagellation, c.* 1460. Urbino, Galleria Nazionale delle Marche

Fig. 1.6 Piero's floor, with a diagonal added

Unlike the unknown illustrator above, Piero allows parallels in the floor to meet on the horizon, which enables him to get the diagonals right. Figure 1.6 shows a close-up of the floor in the picture, with a diagonal superimposed. Notice how the diagonal passes precisely through the corners of tiles. (The contrast has been heightened to show the tiles more clearly.)

In fact, *getting the diagonals right* is the whole secret of drawing a tiled floor in perspective. It is the basis of a method which may be called the *diagonal method*, first appearing in the book *De pictura* (*On painting*) of Leon Battista Alberti in 1436 (Fig. 1.7).

Fig. 1.7 Masaccio: *Portrait of Alberti, c.* 1423–1425. Boston, Gardner Museum

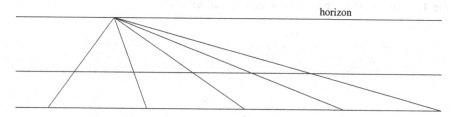

Fig. 1.8 Constructing the first row of tiles

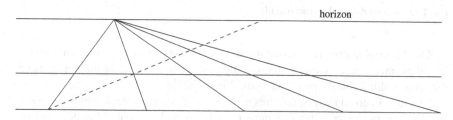

Fig. 1.9 Adding the diagonal

Alberti's diagonal method begins with a series of equally spaced marks on a line parallel to the bottom of the picture (representing the corners of the first row of tiles) and then draws lines from these points to a single point on the horizon—the common "point at infinity" of the columns of tiles. A second line, also parallel to the bottom of the picture, then creates the first row of tiles (Fig. 1.8).

The next, crucial, step is to draw the diagonal of a tile in the first row, shown in Fig. 1.9 as a dashed line.

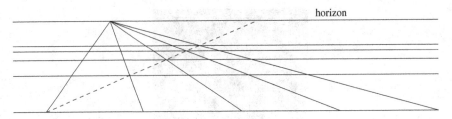

Fig. 1.10 Constructing the second, third, fourth, ... rows

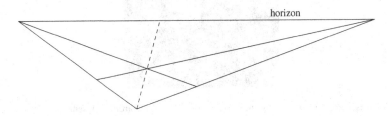

Fig. 1.11 First tile in an arbitrary tiled floor, with diagonal

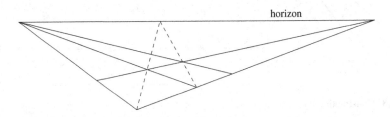

Fig. 1.12 Second diagonal and second tile

The diagonal necessarily passes through the corners of tiles in successive rows, hence its intersections with the lines going to the horizon give us the positions of the second, third, fourth rows, and so on (Fig. 1.10).

Alberti's diagonal method requires a sequence of equally spaced points along the bottom line, and hence it involves measurement. In the language of Euclid, it uses a "compass" as well as a "straightedge."

However, it is possible to produce a perspective view of a tiled floor using a straightedge alone. Also, one can begin with the initial tile in *any* orientation. As in Alberti's diagonal method, everything follows from the diagonal of the first tile (Fig. 1.11).

The diagonal of the second tile is parallel to the diagonal of the first, hence the two meet on the horizon, and this gives us the second tile, and so on (Fig. 1.12).

It is not clear who first drew tiled floors without measurement. Lambert (Fig. 1.13) developed a geometry using straightedge alone. It included several basic perspective constructions, but seemingly *not* the construction of a tiled floor (see Lambert 1773). It may be that the construction of a tiled floor by straightedge

Fig. 1.13 Johann Heinrich Lambert

alone was not contemplated until the nineteenth century, when von Staudt embarked on the construction of addition and multiplication by purely geometric means.

1.2 Projective Planes and Coincidences

In drawing the tiled floor, we seem to be using the following properties of points and lines.

1. There exist four points, no three of which lie in a line. (The vertices of the initial tile.)
2. Through any two points there is a unique line.
3. Any two lines meet in a unique point. (Including parallel lines, which meet at a point on the horizon.)

These are called the *projective plane axioms*. Any collection of objects called "points" and "lines" that satisfy these axioms is called a *projective plane*. The artist's plane, in which there is a "horizon" where parallel lines meet, is modelled mathematically by the so-called *real projective plane*, whose "points" are lines through the origin in three-dimensional space, and whose "lines" are planes through the origin O (Fig. 1.14). One imagines an "all-seeing eye" at O, so the lines through O are "lines of sight." If a picture is drawn on a plane $\mathscr{P}$ not passing through O, there is a unique line of sight through O to each point in the picture. Points in a picture correspond to "points" in the real projective plane. We also have "points" to model the points at infinity of the picture, namely, the lines through O parallel to the plane. In Fig. 1.14 we see three lines through O to points on a line in $\mathscr{P}$. These lines tend towards a parallel to $\mathscr{P}$ as the points move further away.

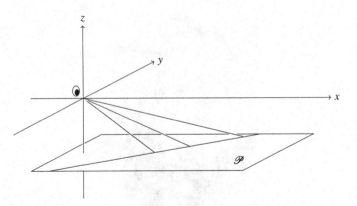

Fig. 1.14 Modelling the projective plane axioms by the real projective plane

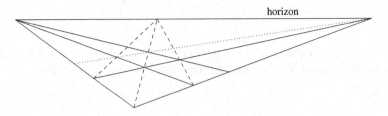

Fig. 1.15 A coincidence in the drawing of the tiled floor

It is easy to check that the real projective plane, which is called $\mathbb{RP}^2$ for short, satisfies the three projective plane axioms. For example, any two "lines" meet in a unique "point" because any two planes through O meet in a line through O. However, $\mathbb{RP}^2$ has some special properties that do *not* follow from the three axioms above.

Sometimes three points fall on the same line for "no reason" (i.e., not as a consequence of the projective plane axioms). I call such an event a *coincidence*. Figure 1.15 shows a coincidence that occurs in drawing the tiled floor: the dotted line passes through three points previously constructed.

In $\mathbb{RP}^2$ there are three famous coincidences, called, respectively:

1. The Pappus theorem.
2. The Desargues theorem.
3. The little Desargues theorem.

These do *not* hold in certain other projective planes, so they may be regarded as additional axioms that "specialize" the plane under discussion. It is known that

$$\text{Pappus} \Rightarrow \text{Desargues} \Rightarrow \text{little Desargues.}$$

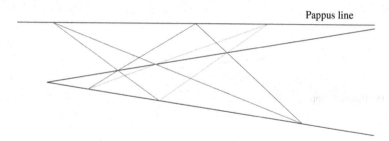

Fig. 1.16 The Pappus configuration

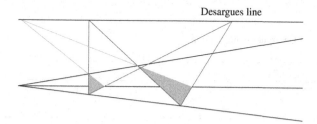

Fig. 1.17 The Desargues configuration

Also,

little Desargues ⟹ all coincidences in the drawing of the tiled floor.

Here is the statement of the Pappus theorem: *for any hexagon with vertices alternately on two lines, the intersections of opposite sides lie on a line*, where the "opposite" sides are the side pairs 1 and 4, 2 and 5, and 3 and 6, respectively. The diagram that illustrates this theorem (Fig. 1.16) is called the *Pappus configuration*. The Pappus theorem appears in his *Collection*, Book VII, from around 300 CE (See Pappus (1986)). It stood alone and hence unrecognised as part of a new kind of geometry until joined by the Desargues theorem over 1,300 years later. Even more remarkable, it was only in the twentieth century that Hessenberg (1905) discovered that the Pappus theorem implies the Desargues theorem

The Desargues theorem states: *for any two triangles in perspective, the intersections of corresponding sides lie on a line*. Triangles are in perspective when the three lines through corresponding vertices have a common point (the "centre of perspective"). The diagram that illustrates this theorem (Fig. 1.17) is called the *Desargues configuration*. Discovered by Desargues in the 1630s, the theorem was first published in the exposition of Desargues' work by Bosse (1648).

In these configurations, the "Pappus line" and "Desargues line" can be the horizon, in which case the pairs of lines that meet there are *parallel*. Figures 1.16 and 1.17 have been drawn with this interpretation in mind. The configurations look

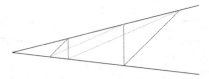

Fig. 1.18 Parallel Pappus configuration

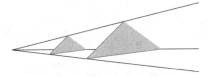

Fig. 1.19 Parallel Desargues configuration

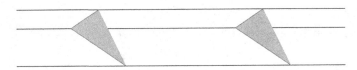

Fig. 1.20 Parallel little Desargues configuration

considerably simpler when the corresponding pairs are drawn to "look parallel," as in Figs. 1.18 and 1.19.

In each case, the theorem takes the form: *if two of the side pairs are parallel, so is the third pair*.

The little Desargues theorem is the special case of the Desargues theorem in which the centre of perspective lies on the line that passes through the intersections of corresponding sides. Thus, if the latter line is taken to be the horizon, the two triangles are in perspective "from infinity" (Fig. 1.20):

Then the little Desargues theorem says: *if two of the side pairs are parallel, so is the third pair*. In $\mathbb{RP}^2$, this theorem is really easy. Nevertheless, it does not follow from the projective plane axioms—there are projective planes in which it does not hold.

1.2.1 The Moulton Plane

Before looking more carefully at the role of the Pappus, Desargues, and little Desargues theorems in geometry, we should take a glance at a plane where none of these theorems hold—the Moulton plane. This plane was devised in 1902 by Forest Ray Moulton, who later became a mathematical astronomer. Apparently he sat in on a course of projective geometry as a student and discovered the plane that now bears his name. The "points" of the Moulton plane are the points of the ordinary

Fig. 1.21 Lines in the Moulton plane

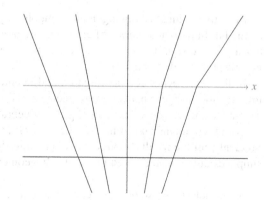

Fig. 1.22 Failure of little Desargues in the Moulton plane

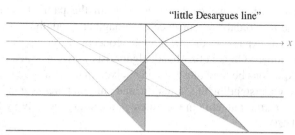

"little Desargues line"

plane $\mathbb{R}^2$, together with its points at infinity, and its "lines" are ordinary lines, except that some of them are bent. Figure 1.21 shows some of these "lines."

The Moulton "lines" are ordinary lines of non-positive slope and "refracted" lines of positive slope—their slope is halved where they pass above the x-axis.

It is not hard to see that these "points" and "lines" satisfy the projective plane axioms (e.g., any two "lines" have a unique "point" in common), but they fail very badly to satisfy even the little Desargues theorem. Figure 1.22 shows why. When all but one of the sides of the triangles are lines of negative slope, the side that bends has an intersection in the wrong place.

It can also be shown that the tiled floor coincidence also fails in the Moulton plane, so this is a projective plane in which Alberti's method for drawing a tiled floor in perspective will break down.

1.3 Geometry and Algebra

Renaissance Italy gave birth not only to a new kind of geometry but also to a new kind of algebra. In the first half of the sixteenth century, Italian mathematicians made the first substantial advance in algebra since ancient times by solving the cubic and quartic equations. These results, due to Scipione Del Ferro, Niccolò Tartaglia, and Lodovico Ferrari, were first published in the *Ars Magna* of Cardano (1545). Despite the revolutionary nature of his material, Cardano treats it very conservatively, interpreting algebraic symbols as geometric entities in the manner of Euclid. In particular, terms of equations are always distributed so that each term has a positive sign, because negative numbers were considered "fictitious" (since negative lengths do not exist).

In the seventeenth century, algebra and geometry advanced together, with the introduction of coordinates into geometry by Fermat and Descartes around 1630. Thanks to the language and technique of algebra, it became possible to describe curves by equations and to investigate their properties by algebraic manipulation. Suddenly, all the results about curves known to the Greeks were demonstrable by simple calculation, and a vast range of new geometric problems became accessible.

Nevertheless, the solution of cubic and quartic equations set algebra on a new path, which eventually diverged from the path of geometry. On the one hand, the unavoidable occurrence of "imaginary" numbers in the solution of cubics led to an expansion of the number concept, to complex (and later, hypercomplex) numbers. On the other hand, algebraists were frustrated by their failure to solve any equations beyond those already solved in the *Ars Magna*. Geometry was of no help in understanding this situation, and enlightenment came only with the development of *abstract* algebra in the nineteenth century, notably by Evariste Galois in the late 1820s.

Galois discovered that the process of solving equations is explained by the *structural* properties of addition and multiplication, such as the *commutative* property of multiplication

$$ab = ba$$

and the *associative* property of multiplication

$$a(bc) = (ab)c.$$

Thus one is led to investigate and classify algebraic structures according to the properties they satisfy. From this point of view, the complex numbers are just another algebraic structure, and indeed one that has the same basic properties as the real numbers. This structural similarity, first anticipated when Bombelli (1572) explained how to reconcile real solutions of cubic equations with the (apparently) imaginary solutions given by the Cardano formula, became an established fact when a rigorous definition of complex numbers was given by Hamilton (1835).

Hamilton constructed the system $\mathbb{C}$ of complex numbers from the system $\mathbb{R}$ of real numbers by forming *pairs* of real numbers that are added and multiplied according to certain rules. The rules are indeed exactly what you get by writing the pair of reals (a, b) as $a + b\mathbf{i}$ and using the rule $\mathbf{i}^2 = -1$ as well as the ordinary rules of algebra. In effect, what Hamilton did was prove the consistency of the rules already used by Bombelli. However, Hamilton's new viewpoint (addition and multiplication of pairs) prompted him to ask: is it possible to add and multiply triples (a, b, c) of real numbers and to satisfy the ordinary rules of algebra? Surprisingly, the answer is no, though it was another 13 years before Hamilton became convinced of this fact.

It turns out that the search for *hypercomplex* numbers—systems of *n*-tuples of real numbers with addition and multiplication that satisfy all (or nearly all) the ordinary rules of algebra—leads to only two systems. They are the *quaternions*, a system of 4-tuples discovered by Hamilton in October 1843, which satisfies all rules except commutative multiplication, and the *octonions*, a system of 8-tuples discovered by John Graves in December 1843, which satisfies all rules except commutative and associative multiplication. We will say more about hypercomplex numbers below, along with a complete listing of the rules they satisfy. Suffice it to say that the solution of the cubic equation in the sixteenth century had surprising consequences in the world of nineteenth century algebra, seeming far from the world of geometry. In the meantime, geometry had also diverged from algebra and number systems. Thanks to the possibility of doing perspective drawings without measurement, projective geometry freed itself completely from coordinates, and became a system in which all theorems were derived by reasoning about points and lines alone. Seemingly, geometry and algebra had diverged completely.

But a surprising development was on the horizon: when geometry is freed from numbers, addition, and multiplication, it becomes feasible to reconstruct algebra on a purely geometric foundation. In the next section we will see how this happened.

1.3.1 Projective Addition and Multiplication

With the introduction of coordinates into geometry by Fermat and Descartes, it became possible to replace geometric arguments by algebraic ones in classical geometry. The same was true in projective geometry. As the name $\mathbb{RP}^2$ suggests, we define the real projective plane in terms of *real numbers*. Lines are defined by equations, and we can prove theorems (such as the Pappus theorem) by computing with numbers and equations.

In his *Geometrie der Lage* of 1847, von Staudt (Fig. 1.23) proposed *doing the reverse*: assuming the Pappus theorem, he defined *addition* and *multiplication* by constructions in projective geometry.

First consider how one might proceed if a compass was available. Given points 0, *a*, and *b* on a line, we can "add *a* to *b*" by opening a compass so that its ends are on 0 and *a*, then sliding the compass parallel to itself along the line until its left point is on *b*, as indicated in Fig. 1.24. The final position of the right compass point is $a + b$.

This construction does not really require a compass. We only need pairs of parallel lines (i.e., lines that meet on some line called the "horizon").

Now notice (Fig. 1.25) that "adding *b* to *a*" to form $b + a$ is a *different construction*:

So, it will be a sheer *coincidence* if $b + a = a + b$. Indeed it is—the Pappus coincidence!—as we see by superimposing the construction of $a + b$ (Fig. 1.26).

Fig. 1.23 Christian von
Staudt

Fig. 1.24 Projective addition
of a to b

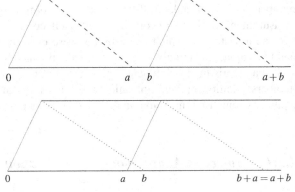

$$0 \qquad\qquad a \quad b \qquad\qquad\qquad a+b$$

Fig. 1.25 Projective addition
of b to a

$$0 \qquad\qquad a \quad b \qquad\qquad b+a=a+b$$

Fig. 1.26 Why
$a+b=b+a$

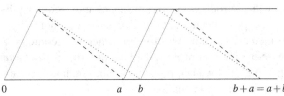

$$0 \qquad\qquad a \quad b \qquad\qquad b+a=a+b$$

There is also a natural projective concept of *multiplication* of points on a line, which looks like *magnification*. First, we open a "compass" to span the points 1 and a on the line. Then we "magnify by b" by sliding the arms of the compass parallel to themselves until the left point is at b. The right point of the "compass" is now ab ("a magnified by b"), as shown in Fig. 1.27.

The construction of ba ("b magnified by a") is different from that of ab, so it again is a *coincidence* if $ba = ab$. But the coincidence happens, thanks to the Pappus coincidence, as one again sees by superimposing the two pictures (Fig. 1.28).

It turns out the Pappus theorem is more crucial for commutative multiplication, $ab = ba$, than it is for commutative addition, $a + b = b + a$. It is possible to derive $a + b = b + a$ from the Desargues theorem, but $ab = ba$ follows only from the Pappus theorem and *not* from the Desargues theorem. As we will see below,

Fig. 1.27 Projective multiplication of a by b

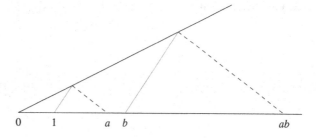

Fig. 1.28 Why $ab = ba$

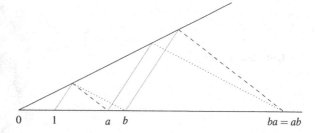

there is a projective plane which satisfies the Desargues theorem but in which ab is sometimes unequal to ba. Thus the commutative law $ab = ba$ gives an *algebraic* sense in which the Pappus theorem is stronger than the Desargues theorem.

There is likewise an algebraic sense in which the Desargues theorem is stronger than the little Desargues theorem. Namely, the associative law $a(bc) = (ab)c$ follows from the Desargues theorem but *not* from the little Desargues theorem. These discoveries, and many related results, are due to Hilbert (1899) and Ruth Moufang (1932, 1933). Moufang (Fig. 1.30) was a mathematical descendent of Hilbert—her Ph.D. supervisor Max Dehn was a Ph.D. student of Hilbert (Fig. 1.29)—and her results are the culmination of the line of research initiated by von Staudt and picked up by Hilbert. Oversimplifying slightly, one may say that von Staudt observed the algebraic role of the Pappus theorem, Hilbert that of the Desargues theorem, and Moufang that of the little Desargues theorem.[1]

In fact, Moufang showed that all of the following *field axioms*, except the two in shades of gray, follow from little Desargues.

$a + b = b + a$	$ab = ba$	(commutative laws)
$a + (b + c) = (a + b) + c$	$a(bc) = (ab)c$	(associative laws)
$a + 0 = a$	$a1 = 1a = a$	(identity laws)
$a + (-a) = 0$	$a^{-1}(ab) = (ba)a^{-1} = b$	(cancellation laws)
$a(b + c) = ab + ac$	$(b + c)a = ba + bc$	(distributive laws)

---

[1] Actually, Moufang stated her results using another theorem, which she erroneously believed equivalent to little Desargues. Hall (1943) pointed out her error and recognised that she had essentially proved results about little Desargues.

Fig. 1.29 David Hilbert

Fig. 1.30 Ruth Moufang

To get $a(bc) = (ab)c$ (associative multiplication) one needs the Desargues theorem to hold. To get $ab = ba$ (commutative multiplication) one needs the Pappus theorem, in which case one gets all field axioms (because Pappus $\Rightarrow$ Desargues $\Rightarrow$ little Desargues). Thus *four geometric axioms (projective plane + Pappus) imply all the field axioms*. (Of which there are nine, because we can drop one distributive law when $ab = ba$.

1.4 Projective Planes and "Number" Systems

We called the plane of perspective drawing the *real* projective plane because real numbers fit into it naturally as *coordinates*, on x- and y-axes.

There is also a natural coordinate m for each point on the horizon; namely, the *slope m* of all lines passing through that point (Fig. 1.31).

With this picture as a guide, we can build the real projective plane $\mathbb{RP}^2$ from three copies of $\mathbb{R}$: the x- and y-axes plus a horizon. Each finite point is an ordered pair (x, y) of reals, and there is a "point at infinity" m for each slope m of a line

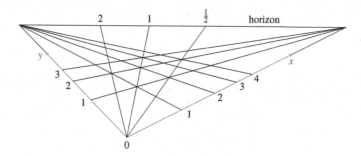

Fig. 1.31 Coordinate axes for the real projective plane

through the origin. Also, the projective constructions of addition and multiplication agree with the addition and multiplication of the real number coordinates.

Similarly, we can build a *complex projective plane* $\mathbb{CP}^2$ from three copies of the *complex number field* $\mathbb{C}$. Since $\mathbb{C}$, like $\mathbb{R}$, satisfies all the field axioms, the Pappus theorem holds in $\mathbb{CP}^2$, and hence also the Desargues and little Desargues theorems.

This raises the questions:

- Can we build a projective plane satisfying Desargues but *not* Pappus?
- Can we build a projective plane satisfying little Desargues but *not* Desargues?

Equivalent algebraic questions to the previous geometric questions are:

- Is there a number system satisfying all the field properties except $ab = ba$?
- Is there a number system satisfying all the field properties except $ab = ba$ and $a(bc) = (ab)c$?

Hilbert gave a (rather artificial) example of a system satisfying all the field axioms except $ab = ba$, but the most natural examples were pointed out by Moufang:

- The *quaternions* $\mathbb{H}$ satisfy all the field axioms except $ab = ba$, so the quaternion projective plane $\mathbb{HP}^2$ satisfies Desargues but not Pappus.
- The *octonions* $\mathbb{O}$ satisfy all the field axioms except $ab = ba$ and $a(bc) = (ab)c$, so the octonion projective plane $\mathbb{OP}^2$ satisfies little Desargues but not Desargues.

The *hypercomplex number systems* $\mathbb{H}$ and $\mathbb{O}$ arise by continuing the "doubling" process that gives us $\mathbb{C}$ from $\mathbb{R}$. $\mathbb{C}$ is a field, consisting of all numbers of the form $a + b\mathbf{i}$, where a and b are real and $\mathbf{i}^2 = -1$.

$\mathbb{H}$ consists of all objects of the form $a + b\mathbf{i} + c\mathbf{j} + d\mathbf{k}$, where a, b, c, d are real and $\mathbf{i}, \mathbf{j}, \mathbf{k}$ are multiplied according to the rules

$$\mathbf{i}^2 = \mathbf{j}^2 = \mathbf{k}^2 = \mathbf{ijk} = -1.$$

$\mathbb{H}$ fails $ab = ba$ because, e.g., $\mathbf{ij} = -\mathbf{ji}$.

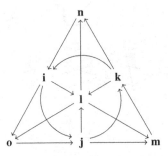

Fig. 1.32 Diagram for octonion multiplication

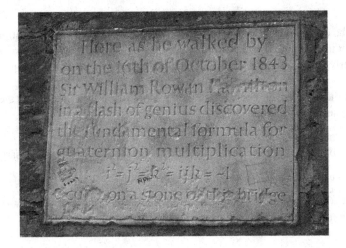

Fig. 1.33 The quaternion plaque on Broom Bridge

$\mathbb{O}$ consists of all objects of the form $a + b\mathbf{i} + c\mathbf{j} + d\mathbf{k} + e\mathbf{l} + f\mathbf{m} + g\mathbf{n} + h\mathbf{o}$, where a, b, c, d, e, f, g, h are real, $\mathbf{i}^2 = \mathbf{j}^2 = \mathbf{k}^2 = \mathbf{l}^2 = \mathbf{m}^2 = \mathbf{n}^2 = \mathbf{o}^2 = -1$, and different units are multiplied according to Fig. 1.32.

The product of one unit by another is the third unit in the same "line" (including the $\mathbf{i}, \mathbf{j}, \mathbf{k}$ circle as a "line"), with a $+$ or $-$ sign according as the step from the first unit to the second is with the arrow or against it. For example, $\mathbf{ij} = \mathbf{k}$, but $\mathbf{ji} = -\mathbf{k}$. $\mathbb{O}$ fails the rule $a(bc) = (ab)c$ because, e.g., $\mathbf{o}(\mathbf{jk}) = -\mathbf{n} = -(\mathbf{oj})\mathbf{k}$.

The quaternion system is denoted by $\mathbb{H}$ in honour of William Rowan Hamilton, who discovered it on October 16, 1843. He discovered the defining equations $\mathbf{i}^2 = \mathbf{j}^2 = \mathbf{k}^2 = \mathbf{ijk} = -1$ while walking across Broom Bridge in Dublin, which now carries a commemorative plaque (Fig. 1.33). The octonions were discovered in December 1843 by Hamilton's friend John Graves. The two had been searching for higher-dimensional analogues of the complex numbers since 1830, until then without success.

Indeed, these were the only such discoveries ever made, because $\mathbb{C}$, $\mathbb{H}$, and $\mathbb{O}$ are the *only* higher-dimensional systems, based on real numbers, that come close to having the same algebraic properties as the real numbers themselves.[2] Thus it is all the more striking that the algebraic properties that *fail* for $\mathbb{H}$ and $\mathbb{O}$— namely, commutative and associative multiplication—should correspond exactly to the Pappus and Desargues theorems of projective geometry. In some sense, there is a natural path leading from perspective drawing to the eighth dimension—but no further.

References

Alberti, L. B. (1436). *Trattato della pittura.* Reprinted in *Il trattato della pittura e i cinque ordine ordine architettonici,* R. Carabba, 1913.

Bombelli, R. (1572). *L'algebra. Prima edizione integrale.* Reprinted by Biblioteca scientifica Feltrinelli, Milano, 1966.

Bosse, A. (1648). *Manière universelle de Mr Desargues.* Paris: P. Des-Hayes.

Cardano, G. (1545). *Ars magna* [The great art or the rules of algebra by T. Richard Witmer, English translation, 1968]. Cambridge, MA: MIT Press.

Ebbinghaus, H.-D., Hermes, H., Hirzebruch, F., Koecher, M., Mainzer, K., & Neukirch, J., et al. (1991). *Numbers.* Berlin: Springer.

Hall, M. (1943). Projective planes. *Transactions of the American Mathematical Society, 54,* 229–277.

Hamilton, W. R. (1935). Theory of conjugate functions, or algebraic couples. Communicated to the Royal Irish Academy, 1 June 1835. In his *Mathematical Papers* (Vol. 3, pp. 76–96).

Hessenberg, G. (1905). Beweis der Desarguesschen Satzes aus dem Pascalschen. *Mathematische Annalen, 61,* 161–172.

Hilbert, D. (1899). *Grundlagen der Geometrie.* Leipzig: Teubner.

Lambert, J. H. (1773). Letter to Karsten, 7 November 1773. In Lambert's *Briefe* (Vol. 4, p. 325).

Moufang, R. (1932). Die Schnittpunktsätze des projektiven speziellen Fünfecksnetzes in ihrer Abhängkeit voneinander. *Mathematische Annalen, 106,* 755–795.

Moufang, R. (1933). Alternativkörper und der Satz vom vollständigen Vierseit. In *Abhandlungen der Math. Seminar* (Vol. 9, pp. 207–222). Hamburg: Hamburg Universität.

Moulton, F. R. (1902). A simple non-Desarguesian plane geometry. *Transactions of the American Mathematical Society, 3,* 192–195.

Pappus. (1986). *Collection. Book VII.* New York: Springer

Savonarola, G. (ca. 1497). *Dell'arte del ben morire.* Florence: Bartolommeo di Libri.

von Staudt, C. (1847). *Geometrie der Lage.* Nürnberg: Bauer und Raspe.

---

[2] See Ebbinghaus et al. (1991), for more on this question.

Chapter 2
Seeing Reality in Perspective: The "Art of Optics" and the "Science of Painting"

Nader El-Bizri

This chapter examines the adaptive assimilation and innovative conceptual prolongations with practical applications of the classical Greek–Arabic science of optics in Renaissance perspectival pictorial arts, as mediated by European mediaeval optical theories and experimentations. This line of inquiry gives a historical account of the epistemic bearings of the connections and distinctions between the exact sciences and the visual arts, with an emphasis on the role of classical optics in the art of painting, and the function of pictorial art in pre-modern natural sciences. A special focus will be set on examining the optical and geometrical legacy of the eleventh century Arab polymath, al-Hasan ibn al-Haytham (known in Latinate renditions of his name as "Alhazen" or "Alhacen"; d. after 1041 CE). This investigation considers the fundamental elements of his theories of vision, light, and space in the context of his studies in optics and geometry, while taking into account his use of experimentation and controlled testing as a method of demonstration and proof. This course of analysis will be furthermore linked to the adaptation of Ibn al-Haytham's research within the thirteenth century Franciscan optical workshops, while scrutinizing the impress that his transmitted texts had on Renaissance perspectival representation of spatial depth and its entailed organization of architectural locales and spaces.

N. El-Bizri (✉)
American University of Beirut, Lebanon
e-mail: nb44@aub.edu.lb

R. Lupacchini and A. Angelini (eds.), *The Art of Science*,
DOI 10.1007/978-3-319-02111-9_2,
© Springer International Publishing Switzerland 2014

2.1 Art of Science and Science of Art

The *dictum*: "*ars sine scientia nihil est*" ("art without knowledge [science] is nothing"), which was attributed to the fourteenth century French architect Jean Mignot (Ackerman 1949),[1] and echoed in Martin Kemp's *The Science of Art* (Kemp 1990), is inverted by Annarita Angelini and Rossella Lupacchini in the thematic orientation of *The Art of Science*, which tacitly asserts that "knowledge [science] without art is nothing" ("*scientia sine arte nihil est*"). This state of affairs situates us within a *liminal* place in-between two propositions that are separated while at the same time being gathered in a dialectical unity: "*l'art n'est rien sans la science' 'la science sans art n'est rien*". The entangled relationships between the exact sciences and the visual and plastic arts date back to ancient times. In the antique epoch, the multivolume *De architectura* (ca. 15 BCE) of the Roman architect and polymath Marcus Vitruvius Pollio constituted one of the early treatises that demonstrated how the various disciplines that formed classical knowledge impacted architectural thinking and the architectonics of place-making.

Arithmetic, geometry, surveying-mensuration, mechanics, optics, astronomy, and natural philosophy were amongst the principal domains of inquiry that influenced pre-modern architecture as a synthesizing field of intellective reflection, and as an applied sphere of practice within material culture, which in itself offered an idealized embodiment of the visual and plastic arts.

In historical and epistemic terms, the entanglement of art with science found some of its most explicit manifestations in the theoretical treatises of Renaissance scholarship, and in the diverse modes of their architectonic and practical applications in the expansion and articulation of material culture. The boundaries that may have separated art from science became creatively blurred in the Renaissance; especially against the background of the deconstruction of the classical Aristotelian physics. What may be pictured as an epistemic or disciplinary "crisis" in classical natural philosophy offered opportunities for the flourishing of artistic and architectural imagination and thinking, wherein art and architecture opened up the horizons of inquiry and the landscapes of curiosity through freer forms of exploration and inventiveness. The successive epochs of the Italian Renaissance were marked by an affirmation of "the *art* of science" and "the *science* of art" at the same time. The scientific grounds of the visual-plastic arts and the artistic underpinnings of the exact-natural sciences were co-entangled. Such dynamics were evident in the context of reflections on the connection and distinction between the *perspectiva naturalis* of visual perception, and the *perspectiva artificialis* of the pictorial representation of the perceptual field of vision. The leitmotifs of *perspectiva* offered an optimal context for investigating the relationships between science and art, in terms of probing the optical and geometric foundations of

---

[1]This *dictum* has been reported in connection with an anecdote about a dispute that took place over the assessment of the structural integrity of the elevation of the *Duomo di Milano* of the Santa Maria Nascente.

the pictorial representation of natural phenomena, while experimenting with the manner painting and drawing in perspective would contribute to the construction of legitimate and reliable knowledge about the visible reality.

When scientific images are radically removed from the familiarities of natural visual perception they necessitate the establishment of complex representational spaces that render the conditions of their observational perceptibility possible. Such modes of picturing reality find their roots in the entanglement of art with science through the course of the unfolding of Renaissance thought and its spheres of praxis. However, if the processes of knowing, proving, and representing are connected with imagining and imaging, does this signal symmetrical relations between science and art instead of asymmetries?

The refinement of representational space, which is pivotal in the enactment of the production of science and art, depended on variegated explorations that were set forth in pursuit of the "*costruzione legittima*" ("legitimate construction") of linear and central single-point perspective within the pictorial art of the Renaissance. Such artistic endeavours were mediated by investigations that also rested on the classical traditions of optics and geometry in the exact sciences. The entanglement of the elements of the pictorial art with the scientific taxonomies in the Renaissance may have been animated at its core by ontological–theological intentions in establishing metaphorical and symbolic connections between scriptural-textual exegesis and the presupposition of visual atonement in measuring reality via the "*visio intellectualis*". Despite the fact that the visual illusory depiction of spatial depth, in the geometric construction and projection of perspective, alluded also to higher orders of "reality", which transcended the way the "real" manifested itself empirically and experientially in visual perception, what concerns us in this line of inquiry is an investigation of the connection and distinction between art and science in relation to the *perspectiva* traditions (El-Bizri 2007a, 2010a,b), and not the explicit reflection on their theological bearings.

The pictorial order is intrinsically implied within the visual elements of the science of optics and of geometry. It also rests on these sciences in the projections and constructions that underpin its representational depiction of spatial depth in artificial perspective.

Examining the visualization of reality and the picturing of the world through the agency of perspective, in terms of natural vision and pictorial representation, constitutes an inquiry into the "art" of optics and the "science" of painting. The epistemic concerns that animated the Renaissance disputations around linear central single-point perspectives in the pictorial arts of the *Trecento*, *Quattrocento*, and *Cinquecento*, all offer concretized historical settings for such line of inquiry as set against the principal theories of the classical sciences of optics and geometry.

To situate this study in a deeper historical *milieu* that underpinned many facets of the *episteme* of Renaissance pictorial arts, I will principally focus on elucidating the key elements of the optical and geometrical legacies of the eleventh century Arab polymath al-Hasan ibn al-Haytham (known in Latinate renderings of his name as "Alhazen" or "Alhacen"; born in Basra ca. 965 CE, and died in Cairo ca. 1041 CE), with a particular focus on the seven books that constituted his monumental optical opus: *Kitab al-Manazir, The Book of Optics* (Ibn al-Haytham 1983, 1989).[2]

Even though Renaissance scholars were more often inclined theoretically to follow Euclid, Ptolemy, and Vitruvius, they nonetheless relied in optics and in selected aspects of geometry on the transmitted traditions that were associated with Ibn al-Haytham as these were adaptively assimilated and mediated by mediaeval European opticians and mathematicians, leading ultimately to the transformation of the *natural visual theory* into a *pictorial theory*. However, before we become directly engaged in an exegetical and hermeneutic interpretation of the *perspectiva* traditions in connection with Ibn al-Haytham's research in optics and geometry, we need still to probe more closely in the following section some of the entailments of the representational space of pictorial art and scientific imaging.

2.2 Representational Space

The epistemic, veridical, and apodictic criteria of *scientia*, as a source of reliable and sound rational knowledge when conducted within the parameters of precision in logical reasoning and experimenting, are not dependent on personal choices, as it is for instance the case with the spheres of theory and praxis in art, which do not necessitate strict rules of proof and demonstration. This liberal aspect in the explorative horizons of the visual and plastic arts opened up new spheres of inquiry that were imaginatively inventive and relatively freed from the need to follow with stricture the principles of scientific logic and its methodological directives. This state of affairs assisted in the constitution of imaginary models of empirical reality through pictorial representational spaces, which themselves offered contexts for informing the spatial and architectonic qualities of actualized physical architectural locales, specifically through the agency of design and its approximation of the realization of its own formal-material hypotheses.

The rigorous rationality that underpins the coherence of representational space in modelling an imaginative reality within the spectacle of linear central perspective is based on an inner geometric system of points, angles, axes, converging lines, and triangles. The representational space of pictorial perspective is imagined, and then depicted afterwards, or in a succession through the structuring order of geometric construction and projection. Such pictorial space is furthermore refined by way of colour and the anatomy of figurative forms of human and living beings, with their gestures and choreographies, which all manifest a virtual new reality that is saturated with communicative visual metaphors and symbolic meanings. These become vital in their turn in terms highlighting the role of imagination in pictorial and figurative representation, and in the un-concealment of hidden physical and mathematical

---

[2]I used simplified transliterations for all the Arabic terms throughout the text without the noting of diacritical vocalizing marks.

principles of reality. The science that grounded the pictorial arts became itself served by the unfolding of their applications, in founding the role of imagery in the scientific modelling of realities that remained otherwise imperceptible in the course of lived experiential and empirical ambient settings of our human sensibility and its sensorial conditions.

The designer or painter–architect contemplates and imagines certain spatial and architectonic possibilities, which belong to reflections on a given pictorial or architectural context and are mediated via concepts that set down the theoretical hypotheses of design. Such processes unfold through conjectures and the exploration of the most probable possibilities by testing them through drawing, drafting, tracing, and in terms of scaled-models, as physical "*maquettes*". These procedures enact calculative, intuitive, and imaginative strategies that attempt to approximate in actualization what can possibly be done in tangible terms within physical reality. The logic of geometry, physics (statics), architectonics, material mechanics, formal, and spatial qualities, atmosphere in imagined sensorial experiences, all bring science and art together in design, while also being oriented by the agency of language in articulating thinking and the manner it depicts the gradual emergence of a composite of form and matter in making. Artistic visions are therefore all along co-entangled with scientific abstractions.

The pictorial representational space that is depicted through artificial linear central perspective makes the seeming sense of infinity manifest in virtual visual terms. The material paintings on the surfaces of canvas appear as windows that are carefully opened up into given regions of imagined worlds, which are chosen through the agencies of the painters and their inherence in history, culture, and language, and are also offered as a complex web of narratives to the observers, be it those who are contemporaneous patrons, or eventually as anonymous spectators that are yet to come in posterity. A human viewpoint on the world is established by *seeing reality in perspective*. A relationship is set between the finite distance of the painter–observer from the surface of the painted canvas, and the implied sense of infinity within the representational virtual space of the depicted portion of imagined reality in the painting.

Two pyramids-cones of visibility intersect in seeing by way of perspective: the finite pyramid-cone of vision of the *perspectiva naturalis*, as studied in optics in connection with direct visual perception, and the pyramid-cone of the *perspectiva artificialis* in the pictorial order, which seemingly tends towards infinity. The pyramid-cone of vision in the *perspectiva naturalis*, as entailed by direct visual perception, is finite and determined by the nearness of its vertex (which is at the centre of the eye of the painter–observer) to its base. As for the pyramid-cone in the *perspectiva artificialis* pictorial order, it gives the semblance of tending towards infinity through the converging geometric lines that meet in the centring-vanishing point on the horizon line. This can be illustrated by the geometric projections in perspective as shown in Fig. 2.1. Let the position of the eye of an observer be seen in a top-view plan as point **O**. Let the lines extended out from this point **O** delimit a cone of vision *CV* that encompasses a given box-shaped object of vision with a vertical surface **a** as it is also seen in a top-view plan. Let *PP* be the picture plane

Fig. 2.1 Geometric
projections in perspective

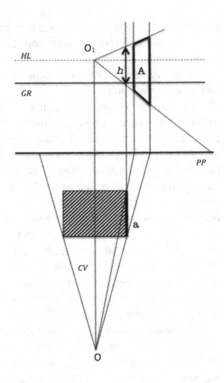

(namely the surface of the canvas) as seen in the top-view plan. Project point **O** into a centring-vanishing point O_1 that appears in a front-view elevation on a horizon line *HL* at the height of the eye of the observer above the ground level *GR*. Let *h* be the height of the surface **a** of the object of vision, which when projected in perspective will be encompassed by the lines extending out from O_1 in such a way that the surface **a** appears in perspective in the shape **A**.

The geometry of the configuration shown in Fig. 2.1 is embedded in the single-point linear and central construct of pictorial perspective, which is established from the viewpoint of a fixed angle of vision, which is determined in the form of a triangle when looking at the cone of vision *CV* in a top-view plan.

The *perspectiva artificialis* is static and marked by fixity, in contrast with the manner the eyes continually move and vibrate in scanning the visual field in the *perspectiva naturalis*. The representational space that is depicted via the *perspectiva artificialis* is itself static and fixed, while opening up to a sense of seeming infinitude. The single-point linear and central construct of pictorial perspective, with the fixity and static quality of its representational order, offer an idealized context for abstractness in geometric space, which is unlike what is brought into appearance within the horizons of natural visual perception. Artificial perspective reveals a symbolic order that is modulated by the exact rules of geometry, and it grants an abstractive viewpoint on what remains hidden from natural sight in the concrete fields of empirical and sensible experience. Artificial perspective lets something "omnipresent" appear, through its geometric order; and yet, there is also the virtual sense by which the painter–observer is also looked at from within the painting when gazing at it.

The contemplation of the painting reveals a virtual viewpoint from a seeming infinity, which looks back at the painter–observer, and is situated at the vertex of the pyramid-cone of the *perspectiva artificialis* within the pictorial space; namely, at the centring-vanishing point where parallels in pictorial-depth tend towards as the seeming "infinite", while meeting in it as geometric lines traced on a two-dimensional surface. As if the painter–observer is also supposedly seen from infinity in a gaze coming from within the painting that remains "*omnivoyant*", given the fixity of the angle of vision in the geometric representational structure of the single-point linear and central pictorial perspective. This outlook is densely expressed in Nicolaus Cusanus's "*Figura paradigmatica*", in his *De coniecturis* (*On conjecture*; ca. 1440 CE), which offers an analysis of two intersecting pyramids, one of light (*lux*), as the *pyramidis lucis*, and the other of shadow (*tenebrae*), as the *pyramidis tenebrarum*, which respectively evoke the ideas of unity and manifoldness. Perspective is posited in this context as a channel of communication between divinities and mortals, "God and man" (Cusanus 1514, 1972; Carman 2007). As if the idealized representational space of pictorial perspective carries also a deeper sense of reality in unveiling the geometric order that grounds and structures the visible universe. In opening up to the infinite, the virtual reality of the painting, as an object of sensible experience, in its materiality as paint-pigments brushed on a canvas surface, becomes itself a portion of a much wider world that is enacted in the pictorial art with its communicative meaningful and symbolic internal complexities.

2.3 Optics

In the earlier sections I advanced various observations concerning the conceptual aspects that emerge from reflecting on the connection and distinction between art and science in relation to the roles played by the debates and the explorations of perspective in the Renaissance artistic and architectural *milieu*. I also signalled the significance that is attributed to the adaptive assimilation and interpretive use of optics, as a science of the *perspectiva naturalis*, in informing the epistemic disputations and technical reflections on the "*costruzione legittima*" of the *perspectiva artificialis*. To situate this inquiry in a setting that entangles the history of science with the history of art and architecture, I will mainly focus in this present section on the fundamental aspects of the optical tradition of Ibn al-Haytham (Alhazen) as primarily embodied in his *Kitab al-Manazir* (*Book of Optics; De aspectibus or Perspectiva*), in view of exploring some of its propositions that are relevant to Renaissance "perspectivism".

The most poignant revolution in the classical science of optics, from the times of Ptolemy to those of Kepler, is embodied in the research of Ibn al-Haytham, who devised a scientific solution to ancient controversies over the nature of vision, light, and colour, which were disputed between the classical mathematicians (exponents of Euclid and Ptolemy) and the Aristotelian physicists. Ibn al-Haytham's research in optics (including his studies in catoptrics and dioptrics, respectively on the principles and instruments of the reflection and refraction of light) also benefited from the investigations of his predecessors in the Archimedean-Apollonian tradition of ninth century Arab polymaths, like the Banu Musa and Thabit ibn Qurra, and of tenth century mathematicians, like al-Quhi, al-Sijzi and Ibn Sahl.[3]

Ibn al-Haytham's *Kitab al-Manazir* was translated from Arabic into Latin towards the end of the twelfth century under the title: *Persepctiva*,[4] or *De Aspectibus*. A fourteenth century Italian version of Ibn al-Haytham's *Optics*, titled: *Prospettiva*, acted as the main reference in optics for the Renaissance sculptor and theorist Lorenzo Ghiberti.

The Latin version of Ibn al-Haytham's *Optics* impacted the research of Franciscan scholars of optics in the thirteenth century, mainly in the 1260s and 1270s, of figures such as Roger Bacon, John Peckham, and Witelo.[5] Ibn al-Haytham's tradition also influenced the investigations of fourteenth century opticians, like Theodoric (Dietrich) of Freiburg (d. ca. 1310) in Europe, and Kamal al-Din al-Farisi (d. ca. 1319 CE) in Persia; both scholars offered correct experimentally oriented explications of the phenomenon of the rainbow and its colouration, while basing

---

[3]While Ibn al-Haytham's optical research proved to be a revolutionizing tradition in the course of development of the scientific discipline of optics up to the seventeenth century, other legacies in this science existed in the history of ideas in the classical Islamic civilization. One of these principal traditions is attributed to the research of the Arab philosopher al-Kindi (d. ca. 873), who partly influenced the optical investigations of Robert Grosseteste (d. ca. 1253) through the Latin version of his treatise in optics, entitled: *De Aspectibus*. However, this optical tradition was primarily Euclidean and Ptolemaic, like it was also later the case with the research of the Persian mathematician and philosopher, Nasir al-Din Tusi (d. ca. 1274). It is also worth noting in this regard that the philosopher and physician Ibn Sina (Avicenna, d. 1037 CE) developed a physical "intromission" theory of vision that is akin to that of Aristotle. Ibn Sina's contributions in optics were not as influential as those of Ibn al-Haytham. Nonetheless, his research on the anatomy of the eye in his *al-Qanun fi al-tibb* (*The Canon of Medicine*) impacted the evolution of ophthalmology up to the sixteenth century, and his research in meteorology inspired Kamal al-Din al-Farisi's revision of Ibn al-Haytham's *Optics* in terms of offering a reformed explication of the reality of colours and the rainbow. Furthermore, Ibn Sina's theory of perception was ecumenically influential in Islamic civilization and European mediaeval scholarship, particularly in terms of elucidating philosophical meditations on the nature of the soul (*al-nafs; De anima*) and the bearings of its cognitive faculties in terms of visual perception (Al-Kindi 1950–53, 1997; Hasse 2000).

[4]The manuscript of the fourteenth century Italian version of Ibn al-Haytham's *Optics*, entitled: *Prospettiva*, is dated on 1341 CE, and it is preserved in the Vatican under the following cataloguing details: Ms. Vat. At. 4595. Folios 1–177.

[5]In a critical analysis of Alistair C. Crombie's thesis that "modern" scientific methodology is attributable to the tradition of Robert Grosseteste, and to thirteenth century opticians like Roger Bacon, John Peckham, and Witelo, Alexandre Koyré argued that the scientific method found its earlier roots in the legacy of Ibn al-Haytham (Alhazen) in optics, which resulted in the flourishing of the perspectivism of Franciscan scholars in the European Middle Ages, in addition to the application of their experimental methods (Koyré 1948; Crombie 1953; Simon 1997; Federici Vescovini 1990, 2008) .

their studies on reformed revisions of Ibn al-Haytham's theory of colours as noted in his *Optics* (Federici Vescovini 1990, 2008).[6] For instance, Kamal al-Din al-Farisi conducted an experiment on a large spherical glass vessel modelling a rain-droplet, which was subjected to light in a controlled environment within a *camera obscura* (*al-bayt al-muzlim*), to demonstrate the decomposition of white light into a spectrum of colours, in view of explicating the phenomenon of the rainbow in meteorological optics (El-Bizri 2009). Ibn al-Haytham's tradition in history of science in Islam continued to be subsequently influential through the investigations of the Syrian astronomer at the Ottoman court, Taqi al-Din Muhammad Ibn Ma'ruf (d. ca. 1585 CE; El-Bizri 2005a).

The Latin translations of Ibn al-Haytham's *Kitab al-Manazir*, in addition to works associated with geometry and conics, in relation to Arabic sources in mathematics, also impacted Renaissance scholars of the calibre of Biagio Pelacani da Parma (Pelacani da Parma 2002),[7] Francesco Maurolico, Ettore Ausonio, Egnatio Danti, and Francesco Barozzi.[8] Ibn al-Haytham's *Optics* was also assimilated in Renaissance scholarly circles, partly through the mediation of thirteenth century Franciscan opticians, and it influenced the perspective theories of Leon Battista Alberti in the *De pictura*, and impacted more directly the propositions of Lorenzo Ghiberti in the *Commentario terzo* (Federici Vescovini 1998). A printed edition of Ibn al-Haytham's Latin version of the Optics was established by Friedrich Risner in 1572 in Basle, under the title: *Opticae Thesaurus*, which was eventually consulted by seventeenth century scientists and philosophers such as Kepler, Descartes, Huygens, and possibly even Newton. The recognition of Ibn al-Haytham's *œuvre* is also evident in the high station he was accorded by the seventeenth century German scientist Johannis Hevelius, whereby the frontispiece of the latter's *Selenographia sive Lunae Descriptio* (dated 1647) depicts Ibn al-Haytham standing on the pedestal of *ratione* (reason), with a compass in his hand and a folio of geometry, while Galileo stands on the pedestal of *sensu* (observation), holding a telescope.

An investigation of the historical and epistemic entailments of Ibn al-Haytham's tradition in optics elucidates some of the dynamics that are at work in the emergence and development of novel scientific rationalities. His legacy established the principal scientific foundations of mediaeval *perspectiva* in the European traditions, and, through them, it grounded in part selected Renaissance theories of vision and

---

[6]Phenomena that were originally treated as topics of meteorology were studied based on new models of "reformed" optics. For instance, Kamal al-Din al-Farisi's (d. ca. 1319 CE) explication of the phenomenon of the rainbow (*qaws quzah*) constituted a part of his commentary on Ibn al-Haytham's *Optics* in *Tanqih al-manazir*; namely, a treatise entitled: *The Revision of [Ibn al-Haytham's] Optics* (Al-Farisi 1928–29).

[7]This is particularly the case with the *Quaestiones perspectivae* of Biagio Pelacani da Parma.

[8]He is also known as "Franciscus Barocius", and this particular discussion figures mainly in his *Admirandum illud Geometricum Problema tredecim modis demonstratum*—Raynaud and Rose discussed some related elements of the adaptive assimilation by Renaissance theorists of Arabic mathematical sources on conics and their applications in optics (Raynaud 2007; Rose 1970).

perspective, while continuing furthermore to influence the unfolding of the science of optics up to the seventeenth century.

Ibn al-Haytham's scientific method consisted of combining mathematics with physics in the context of experimental demonstration, verification, proof, and controlled testing (*i'tibar muharrar*), including the design and use of scientific instruments and installations (El-Bizri 2005a). Ibn al-Haytham investigated the veridical conditions of visual perception to ground the observational data of his experimental research, along with setting rigorous parameters for the application of optics in astronomy and meteorology.

One of the principal aspects of Ibn al-Haytham's reforming of the science of optics is encountered in his ingenious resolution of the longstanding ancient dispute between the mathematicians (*ashab al-ta'alim*; Euclidean and Ptolemaic) and the physicists (*ashab al-'ilm al-tabi'i*; Aristotelian) over the nature of vision and light. Ibn al-Haytham showed that vision occurs by way of the introduction of physical light rays into the eye in a configuration that is geometrically determined in the form of a pyramid-cone (*makhrut*) of vision, with its vertex at the centre of the eye and its base on the visible lit surfaces of the object of vision; while taking into account the rectilinear propagation of light in the homogeneous transparent medium between the observer and the seen object. He thus rejected the "extramission" theory of the ancient mathematicians, which holds that vision occurs by way of the emission of a subtle and non-consuming ray of light (akin to fire) from the eye that meets the lit medium, which, as a physical phenomenon, is structured in the form of an actual pyramid-cone of light. In view of explicating the process of vision, Ibn al-Haytham retains the structural form of a pyramid-cone of vision, in terms of geometric modelling, while emphasizing that it is abstracted from matter, and that the lines determining its outline and configuration were purely mathematical (virtual and postulated) rather than being physical. Moreover, he refuted the physicists' theory of vision (as inspired by Aristotle's *Physics* and *De anima*), which ambivalently conjectured that the sight results from the "intromission" into the eye of the form of the visible object without its matter when the transparent medium (*al-shafif; diaphanes*) is actualized by physical illumination. Ibn al-Haytham demonstrated that vision occurs by way of the introduction of light into the eye, while showing that this physical phenomenon was geometrically structured in the shape of a virtual-mathematical cone of vision (Nazif 1942–43; Federici Vescovini 1965; Sabra 1978, 1989; Rashed 1992). Consequently, he distinguished vision from light, and devised novel methodological procedures that brought the certitude and invariance of geometrical demonstration to bear with isomorphism instead of mere synthesis on his research in physical optics (El-Bizri 2005b). He moreover subjected the resultant mathematical-physical models and hypotheses to experimentation by way of controlled empirical procedures of testing, including the devising and use of experimental instruments and installations, like the *camera obscura* (Nazif 1942–43; Schramm 1963; Omar 1977). Moreover, his experimentation did not consist of a simple element of empirical methodology, rather it was theoretically integral to his proofs, and granted an apodictic value to his inquiries in optics (Rashed 2005; El-Bizri 2005b).

Ibn al-Haytham's geometrical, physical, physiological and meteorological studies in optics were also related to his psychology of visual perception, and to his analysis of the faculties of judgement and discernment (*al-tamyiz*), of cognitive comparative measure (*al-qiyas*), of (*eidetic*) recognition (*al-ma'rifa*), imagination and memory (*al-takhayyul, al-dhakira*). He thus distinguished the *immediate* mode of perception by way of glancing from *contemplative* perception (*Optics*, II.4: 5, 20, 33)[9] while reflecting on the manner of perceiving particular visible properties (*al-ma'ani al-mubsara; intentiones visibiles—Optics*, II.3: 43–48).[10] Pure sensation only perceives light *qua* light and colour *qua* colour (*Optics*, II.3: 50–52; II.4: 22), while vision depends primarily on exercising the *virtus distinctiva* (*al-quwwa al-mumayyiza*; faculty of discernment), which perceives all visible 22 properties (*Optics*, II.3: 1–25), while being aided by imagination and memory, and usually operating without deliberate and excessive effort (*Optics*, II.4: 12–15, 22). Ultimately, the light introduced into the eyes results neurologically and physiologically in sensations in the last sentient (*al-hass al-akhir; sentiens ultimum*) in the frontal part of the brain (*muqaddam al-dimagh; Optics*, I.6: 74).

Ibn al-Haytham's observations rested on anatomical examinations of the structure of the eye (*Optics*, I.5: 1–39) and the investigation of binocular vision (*Optics*, I.6: 69–82). "*Why do we see a single object of vision instead of two, even though we look at it with two eyes?*" The image formed on the crystalline of the eye (*al-jalidiyya*) passes through the vitreous ocular humour (*al-zujajiyya*) and reaches the hollow optic nerve (*al-'asaba al-jawfa'*), which connects to the common nerve (*al-'asaba al-mushtaraka*; optic *chiasma*) and reaches the last sentient as a sensation in the anterior part of the brain. Under normal conditions of binocular vision, the observer perceives a single visible object with two sound eyes instead of having two images of one and the same object. Binocular vision does not readily result in double vision, unless this is due to errors in vision (which Ibn al-Haytham examined in detail in Book III of his *Optics*). The form of a single visible object occurs on the surface of the crystalline of each of the eyes. Looking at that object, its form is received in each one of the eyes. This can be illustrated as shown in Fig. 2.2. Let a given point O on the object of vision, respectively, reach the right and left eyes in points O_1 and O_2 that become united into a fused "O_{image}" via the common optical nerve. Consequently, two forms (such as those entailed by points like O_1 and O_2), occur on the crystalline of each of the eyes, passing via the vitreous to the hollow nerves, and then, as sensations, become unified in the common nerve, and reach the last sentient as an ordered single form of a sensible visible object (namely as an "O_{image}").

---

[9]References that are hereinafter made to Ibn al-Haytham's *Optics* in the body of the text indicate the numbering of the Book with its chapters, as these correspond with the Arabic critical edition of the text (Ibn al-Haytham 1983) and its annotated English translation (Ibn al-Haytham 1989).

[10]Ibn al-Haytham enumerated twenty-two particular visible properties (*Optics*, II.3: 44), while Ptolemy restricted their number to seven (Lejeune 1948; Sabra 1966).

Fig. 2.2 The form of a single
visible object occurs on the
surface of the crystalline of
each of the eyes

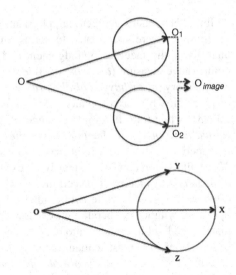

Fig. 2.3 Only the light rays
that meet the outer surface of
the crystalline humour
perpendicularly are admitted
into the eye

Apart from binocular vision, "*why a single object of vision appears as one and not many? And how do multiple light rays, which result in manifold visual data, get received into the eye in an ordered structure?*" The object of vision is seen by way of the introduction into the eye of light rays that are emitted from its visible lit surfaces, which propagate rectilinearly across the transparent medium that is between the eyes of the observer and this object, while the reception of these light rays in the eye is structured geometrically in the shape of a *virtual* cone of vision (*makhrut al-shuʿa*), with its vertex at the centre of the eye and its base on the seen and lit surfaces of the visible object (*Optics*, I.2, I.3). The light rays that are structured within this mathematical model travel rectilinearly from every point on the lit and appearing surfaces of the visible object in a punctiform-corpuscular configuration, with a spherical irradiation that is emitted in rectilinear trajectories from each of these points through the transparent medium, and in all directions. This phenomenon reflects a point-by-point correspondence between each point on the lit and visible surface of the object of vision and each correlative point on its image that occurs on the crystalline, which ultimately secures the ordering of the visible aspects of this seen object. Only the light rays that meet the outer surface of the crystalline humour (*al-rutuba al-jalidiyya*) perpendicularly (centrally at a normal) are admitted into the eye in terms of this point-by-point correspondence between the lit and visible surfaces of the object of vision and the image they have on the crystalline. This is shown in the Fig. 2.3, whereby a hypothetically given "point" **O** on the object vision will have its corresponding image admitted within the eye as a single hypothetically given "point" **X**.

This phenomenon was also analysed by Ibn al-Haytham in terms of studying the geometrical properties of the outer surface of the crystalline as an optical lens (spherical section) in dioptrics, as set in Book VII of his *Optics*, which focused on the mathematical properties of refractive surfaces (with differing indices of refraction), and derived from varied spherical, cylindrical, and conical sections (circle, ellipsis, parabola, hyperbola). This line of study was partly based on the tenth century research of Abu al-'Ala' Ibn Sahl on geometrically modelled lenses, mainly the latter's *Kitab al-Harraqat (Burning Instruments)*.[11]

The impressions of the luminous physical rays on each of the eyes in binocular vision are ultimately interpreted in the brain as a single integral image of one object of vision, which is not the same as a mental image *per se*. On Ibn al-Haytham's view, vision is a physiological, neurological, and psychological/cognitive phenomenon that is not simply reducible to the order of geometric and physical analytics in optics. Vision is investigated in this regard from the standpoint of the epistemic dimensions of cognition, and not solely from the standpoint of mathematical-physical models. This nuance indicates the possibility of translating his theory of vision into two forms of "visual cultures": one that imitates the visible realm in "pictorial representations", and the other conducts thinking through the agency of "mental pictures". Respectively, these correlate with the representational spaces of art and science.

2.4 Renaissance Perspectives

The Renaissance perspective geometric constructs aimed at reproducing with two-dimensional approximation and pictorial reinterpretation the three-dimensional spatial spectacle that is naturally perceived by eyesight. This procedure rested on the science of optics, though finding novel spheres of its application and significance in the context of the visual fine arts. Objects of vision are to be depicted as they appear. This effort started with *Trecento* artists in terms of representing what we see in pictorial terms, without yet having developed a rigorous geometric method to construct linear perspective. Even though such aspects of depicting things as they appear may have had some much earlier manifestations in examples from antiquity (including murals in some of the Roman villas in Pompeii), the preoccupations of the *Trecento* painters were addressed via conscious studies rooted in the classical science of optics.

Geometry was used in the science of optics, within the tradition of Ibn al-Haytham, in isomorphism with physics and controlled experimentation. This endeavour aimed at scientifically explaining the nature of visual perception, and the laws of the rectilinear propagation of light in a homogeneous transparent medium, the reflection of light on polished surfaces (catoptrics), and the refraction of light when passing from a transparent medium into another that differs from it in subtlety and in its refractive indexing (dioptrics). In the case of art and architecture, geometrical constructions and projections eventually acted as tools for the depiction

---

[11]This aspect had implications on studying spherical aberration; namely, when beams of light, which are parallel to the axis of the lens (as a spherical section), yet that also vary in terms of their distance from it, become all focused in different places, which results in the blurring of the resultant image.

of spatial depth in pictorial representations, and served also as design directives in the organization of architectural space and the articulation of its architectonic features in concrete physical settings.

In *Le due Regole della prospettiva*, first published in 1583, Jacopo Barozzi (known as Vignola) dedicated a chapter to refute the idea of constructing linear perspective through two vanishing points that correspond with binocular vision (Barozzi 1611). In this, he deployed arguments that accorded with what Ibn al-Haytham demonstrated in terms of the psychological–neurological–physiological aspects of vision, by way of accounting for binary visual perception, and the fusion-unification of the visible form of the object of vision, when the light rays emitted from the visible lit surfaces of that object make their final impress, via the eyes and the optical nerves, on the last sentient located in the anterior part of the brain (principally as noted in Chap. 6 of Book I of the *Optics*). This aspect of binocular vision, and its implications in terms of thinking about the method of constructing linear perspective in pictorial representational art, attracted also the comments of the Renaissance mathematician Egnatio Danti who sustained similar views as those of Jacopo Barozzi, as he commented on the latter's opus (Raynaud 2003, 2004). Danti displayed also signs of awareness with regard to these observations in the science of optics, and in lines that accorded with Ibn al-Haytham's theories. Ultimately, linear perspective is said to have a single centring-vanishing point instead of two, hence, being mono-focal and central, without contradicting the nature of binocular vision. However, the traditions practiced in the *Trecento* pictorial renderings, based on asserting binocular vision, tended to posit two vanishing points that are correlative with the two eyes of the observer, without being in this "bifocal" in the sense of having a two-point perspective that is associated with relatively more modern constructs (like the ones that are also "trifocal", or "curvilinear", etc.). Notwithstanding, the science of optics, as exemplified by Ibn al-Haytham's theory of visual perception, and his analysis of binocular vision, allowed for two pictorial interpretations: the *first* consists of positing a single centring-vanishing point in mono-focal central linear perspective, which correlates with the presupposition of a single cone of vision receiving the seen spectacle by the observer, and taking into account the fusion of impressions on the eyes through the common optical nerve (as discussed earlier and shown in Fig. 2.2), while the *second* allows the positing of two vanishing points in asserting binocular vision. The latter was exemplified in what we may call: "the heterodox [*Trecento*] perspectives", which posits two vanishing points; like, for instance it was the case with Lorenzo Ghiberti's *Christ Amongst the Doctors* (fourth panel of the North door at the Baptistery of San Giovanni).

The *problématique* of the "*costruzione legittima*" (legitimate construction of perspective) centred on the consequences of doubling the unique centring-vanishing point of central perspective, and on debating the risks of distortions, or of compromising the spatial unity of the representational pictorial field. The manipulation of heterodox two-point perspectives, in terms of depicting central foreground figures against architectural background settings, to neutralize the effects of diplopia, did not always succeed in avoiding visual distortions, or in securing the unity of the painted representational space (Raynaud 2004).

The pictorial interpretation based on the heterodox positing of two-vanishing points (like it was the case with Gentile de Fabriano's *The Tomb of Saint Nicholas*), reflects an optical awareness of the need to accommodate binocular vision instead of monocular sight. However, this consciousness does not account for the fusional convergence of the two images formed on the crystalline of the eyes, and their unification in terms of the physiological-neurological-psychological determinants of vision, as analysed by Ibn al-Haytham. Rather, this practice rests on an analysis of binocular vision that attempts to overcome the effects of double vision, diplopia and parallax phenomena, under normal physiological conditions of eyesight. Such dimensions were also carefully studied in Ibn al-Haytham's optics, in terms of investigating the implications of distance in vision (nearness to the eyes in particular), and of optical convergence or its insufficiency, of visual alignments and misalignments, of parallax phenomena and stereopsis, with their various effects on the positioning of the eyes and the physiological-ocular effort in focusing sight on certain objects within a given spectacle, with the potential also of generating errors in visual perception (principally as studied in Chap. 2 of Book III of Ibn al-Haytham's *Optics*).

Binocular diplopia, commonly known as "double vision", entails the simultaneous perception of two quasi-displaced-images of a single object, which results from the misalignment of the two eyes relative to one another, and due to "convergence insufficiency". This is not an ocular disorder when the object of vision is brought at a near distance to the eyes and results from normal physiological conditions of optical convergence, which require additional effort in focusing the two eyes on an object that is very close to them, and seeing it against the background of other more distant objects. Binocular vision is normally accompanied by *singleness in vision or binocular fusion*, in which one and a single image is seen despite each eye having its own image of the object of vision. Moreover, stereopsis exploits the parallax, in terms of the displacement of a single object viewed via two different lines of sight of the eyes, along with the binocular fusion of these two resultant images, leading ultimately to seeing spatial depth.

The theoretical presuppositions guiding the construction of mono-focal linear perspective are grounded on a sound optical analysis of binocular vision and the singleness of vision in terms of binocular fusion, as analysed by Ibn al-Haytham. This relies on the psychological, physiological, and neurological determinants of visual perception, which result, under normal conditions of vision, in the fusion of two disparate images-forms of a single visible object, as received by each of the eyes of the observer, and being unified in the brain through the agency of the optical nerves, the common optic *chiasma*, and the exercising of cognition in effecting sight.

Alberti and Ghiberti animated the discussions concerning these optical directives in terms of how they underpinned the legitimate methods of constructing perspective, as these embodied varying levels of adapting Ibn al-Haytham's optical legacy and its reception by mediaeval and Renaissance perspectivists. These elements of debate also continued to preoccupy figures such as Piero Della Francesca in his *De Prospectiva Pingendi*, with applications in his *Flagellation* painting (see Fig. 1.5 in chapter 1) that rendered it exemplary amongst the perfected perspectival constructs. Celebrated linear perspectives were also associated with Masaccio's

Santa Trinita (in Santa Maria Novella, Firenze), Donatello's *Banquet of Herod*, or Raphael's remarkable *Scuola di Atene* (in the *Stanza della Segnatura*, Vatican apostolic palace). Investigations that focused on the perfection of the depicted pictorial representational spaces, in the projections and constructions of linear central perspectives, combined with in-depth studies in geometric optics, resulted eventually in perspectival approaches to Euclidean geometry, which culminated in the seventeenth century in advanced legacies of geometric perspectivism, as for instance embodied in Girard Desargues' *Œuvres mathématiques*, and in his projective geometry (Desargues 1647).

In order to grasp the invention of linear perspective based on optics, it is vital to also consider novel ways of accounting for "the visibility of space" since the times of Ibn al-Haytham, by way of considering the mediated reception of prolongations of his legacy in Renaissance circles, and the definition of representational space with an ordered geometry of its own that unifies its imagined visual field. The novel reorientation of the development of perspective in relation to the spatial order of the seen spectacle that is depicted rested on a newly defined "looking space" (Vesely 2004), which required the controlled counterbalancing of the problematic aspects of visual illusions or errors via geometrical structuring measures that facilitated the location of objects and their interrelations within the visual field. This ushered a new phase in the debate over orthogonals, viewing points, vanishing points, and the visual cone-pyramid, which became foundational concepts for the invention of geometrized perspective and its presupposition of a "mathematized space". The idea of perspective as a pictorial representational construct rested on the fundamental notions of a "geometrized space" and "the visibility of spatial depth" (both rooted in Ibn al-Haytham's mathematical and optical research, as we shall highlight in the following section below).

2.5   Geometrical Place as Spatial Extension

Ibn al-Haytham presented his geometrical conception of place as a solution to a long-standing problem that remained philosophically unresolved, which, to our knowledge, also constituted the first viable attempt to geometrize "place" in history of science. This corresponded with Ibn al-Haytham's foundational endeavour to "mathematize physics" in the context of experimental research in optics. Ibn al-Haytham aimed at promoting a geometrical conception of place that is akin to *spatial* extension in view of addressing selected mathematical problems that resulted from the unprecedented developments in geometrical transformations (similitude, translation, homothety, affinity, etc.), the introduction of motion in geometry, the anaclastic research in conics and dioptrics in the Apollonian-Archimedean Arabic legacy since the ninth century (El-Bizri 2004, 2007b).

Besides the *penchant* to offer mathematical solutions to problems in theoretical philosophy that were challenged by longstanding historical obstacles and epistemic impasses, Ibn al-Haytham's endeavour in geometrizing place was undertaken in view of sustaining and grounding his research in mathematical analysis and synthesis (*Fi al-tahlil wa-al-tarkib*),[12] and in response to the needs associated with the unfurling of his studies on knowable mathematical entities (*Fi al-ma'lumat*),[13] and in order to reorganize most of the notions of geometry and rethinking them anew in terms of motion (*al-haraka, al-naql*). Consequently, he had to critically reassess the dominant philosophical conceptions of place in his age, which were encumbered by inconclusive theoretical disputes over Aristotle's *Physics* (Aristotle 1936).

Even though Aristotle affirmed that *topos* has the three dimensions of length, width and depth (*Physics*, IV, 209a 5), he defined *topos* as: "the innermost primary surface-boundary of the containing body that is at rest, and is in contact with the outermost surface of the mobile contained body" (*Physics*, IV, 212a 20–21). Contesting this long-standing Aristotelian *physical* conception of *topos*, Ibn al-Haytham posited *al-makan* as "imagined void" (*khala' mutakhayyal*; postulated void) whose existence is secured in the imagination (like it is the case with invariable geometrical entities). He moreover held that the "imagined void" *qua* "geometrized place" consisted of imagined immaterial distances that are between the opposite points of the surfaces surrounding it (Rashed 1993, 2002). He furthermore noted that the imagined distances of a given body, and those of its containing place, get superposed and united in such a way that they become the same distances (*qua* dimensions) as mathematical lines having lengths without widths-breadths.

From a philosophical viewpoint, we could say that Ibn al-Haytham's geometrical determination of place was "ontologically" neutral. This is the case given that his mathematical notion of *al-makan* was not simply obtained through a "theory of abstraction" as such, nor was it derived by way of a "doctrine of forms", nor was it grasped as being the (phenomenal) "object" of "immediate experience" or "common sense". It is rather the case that his geometrized place resulted from a mathematical isometric "bijection" function between two sets of relations or distances (El-Bizri 2007b).[14] Nothing is thus retained of the properties of a body other than *extension*, which consists of mathematical distances that underlie the geometrical and formal conception of place (Rashed 2002).

To give an example of Ibn al-Haytham's mathematical refutation of Aristotle's physical definition of *topos*, we could consider the case of his geometric demonstration based on the properties of a parallelepiped (*mutawazi al-sutuh*; a geometric solid bound by six parallelograms; a cuboid). If this given parallelepiped were to be divided by a rectilinear plane that is parallel to one of its surfaces, and is then recomposed, the cumulative size of its parts would be equal to its original

---

[12]The Arabic critical edition (based on four manuscripts) and the annotated French translation of this treatise (*Fi al-tahlil wa-al-tarkib; L'Analyse et la synthèse*) are established in Rashed (2002, pp. 230–391).

[13]The Arabic critical edition (based on two manuscripts) and annotated French translation of this treatise (*Fi al-ma'lumat; Les connus*) are established in Rashed (2002, pp. 444–583).

[14]"Bijection" refers to an equivalence relation or function of mathematical transformation that is both an "injection" ("one-to-one" correspondence) and "surjection" (designated in mathematical terms also as: "*onto*'") between two sets.

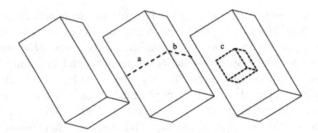

Fig. 2.4 The magnitude of a parallelepiped divided along the lines a and b would increase in surface area by a quantity equal to $2ab$; the magnitude of the same parallelepiped carved out of a cube with a side c would increase in surface-area by a quantity $4c^2$, whilst it would decrease in volume

magnitude prior to being divided, while the total sum of the surface areas of its parts would be greater than its surface-area prior to being partitioned. Following the Aristotelian definition of *topos*, and in reference to this divided parallelepiped, one would conclude that: an object divided into two parts occupies a place that is larger than the one it occupied prior to its division, since its total surface area increased with its division. Hence, the magnitude of the place of a given body increases while the size of that body does not; consequently: "objects of equal magnitudes are contained in unequal places", which is an untenable proposition (Rashed 2002; El-Bizri 2007b). Likewise, if we consider the case of a parallelepiped that is carved, then, its bodily magnitude is diminished while the total sum of its surface area would increase. Following the Aristotelian definition of *topos*, and in reference to this carved parallelepiped, one would conclude that: an object that diminishes in magnitude occupies a larger place, which is untenable.

For example, as shown in the Fig. 2.4, the magnitude of the middle parallelepiped that has been divided along the lines a and b would increase in surface area by a quantity equal to $2ab$. As for the carved parallelepiped to the right side in the figure (Fig. 2.4), if a cube with a side c were to be cut out from it, then its magnitude would decrease, whilst its surface-area increases by a quantity $4c^2$.

Moreover, using mathematical demonstrations, in terms of geometrical solids of equal surface-areas (isepiphanic), and figures that have equal perimeters (isoperimetric), Ibn al-Haytham showed that the sphere is the largest in (volumetric) size with respect to all other primary solids that have equal surface-areas (*al-kura a'zam al-ashkal al-lati ihatatuha mutasawiya*). So, if a given sphere has the same surface-area as a given cylinder, then they occupy equal places according to Aristotle, and yet, the sphere would have a larger (volumetric) magnitude than the cylinder; hence unequal objects occupy equal places, which is not the case.

Ultimately, Ibn al-Haytham's critique of Aristotle's definition of *topos*, and his own geometrical positing of *al-makan* as an "imagined void" (*khala' mutakhayyal*), both substituted the grasping of the body as being a totality bound by physical surfaces to construing it as a set of mathematical points that are joined by geometrical line-segments. Hence, the qualities of a body are posited as an *extension* that consists of mathematical lines, which are invariable in magnitude and position, and that connect points within a region of the *three-dimensional space* independently of the physical body.

The geometrical place of a given object is posited as a "metric" of a region of the so-called "Euclidean" *qua* "geometrical *space*", which is occupied by a given body that is in its turn also conceived extensionally, and corresponds with its geometrical place by way of "isometric bijection". Consequently, Ibn al-Haytham's geometrical determination of place points to what later was embodied in the conception of the "anteriority of spatiality" over the demarcation of a metric of its regions by means of mathematical lines and points, as explicitly implied by the notion of a "Cartesian space" (Rashed 2002; El-Bizri 2007b). The scientific and mathematical significance of the geometrization of place was confirmed through the unfolding of mathematics and physics in seventeenth century conceptions of place as extension (namely as a volumetric, three-dimensional, uniform, isotropic and homogeneous space), particularly in reference to Descartes' *extensio* and Leibniz's *analysis situs*, and the emergence of what came to be known in periods following Ibn al-Haytham's age as being the "Euclidean space" (namely, an appellation that is coined in relatively modern times, and describes a notion that is historically posterior to the geometry of figures as embodied in Euclid's *Stoikheia* [*The Elements; Kitab Uqlidis fi al-Usul*]).[15]

Ibn al-Haytham's reflections on the notion of space in his *Kitab al-Manazir* (*Optics*) were commensurable with his mathematical conception of place in his *Qawl fi al-makan* (*Discourse on Place*). Ibn al-Haytham asserted that spatial depth is a visible property (unlike the eighteenth century immaterialism of George Berkeley, who denied the visibility of space).[16] Ibn al-Haytham also argued that: in order that the distance, which separates the observer from the object of vision, gets estimated, the thing being perceived ought to be near objects that are ordered and contiguous (*Optics*, II.3: 76–80), as well as share a common unified terrain with the observer.

To demonstrate this situational and phenomenological condition, Ibn al-Haytham established an experimental installation that consisted of a wall dividing a given hall into two distinct spaces S_1 and S_2 (as shown in Fig. 2.5), which are visually linked through a pinhole aperture **a** (*thuqb*), piercing the wall separating them, in such a way that the floor and ceiling in space S_1 could not be seen when looking through [a] from S_2. The concealed space S_1 receives objects that could only be viewed by observers in this experiment from S_2 through aperture [a]. If two screen-walls **w1**

---

[15]After all, the expression deployed by Euclid that is closest to a notion of "*space*" as denoted by the Greek term: "*khôra*", is the appellation: "*khôrion*", which designates "an area enclosed within the perimeter of a specific geometric abstract figure", as for instance noted in Euclid's *Data* (*Dedomena; al-Mu'tayat*) Proposition 55 (as also related to: *Elements*, VI, Proposition 25): "if an *area* [*khôrion*] be given in form and in magnitude, its sides will also be given in magnitude" (Euclid 1956, 1883–1916).

[16]This question preoccupied Maurice Merleau-Ponty in the twentieth century, in terms of re-affirming the visibility of spatial depth in his *Phénoménologie de la Perception* (Merleau-Ponty 1945; El-Bizri 2004).

Fig. 2.5 Experimental
installation conceived by Ibn
al-Haytham

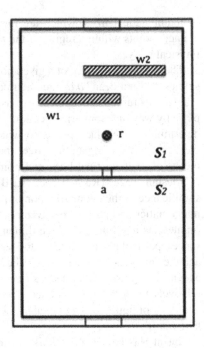

and **w2** were to be introduced into the concealed space S_1 at different distances from
the dividing wall, then, looking through the aperture [a], the observers in space S_2
could not detect the difference between the distances of the screen-walls **w1** and
w2; and when these screen-walls were subjected to an intense light, the observers
were not able to even distinguish them from each other (*Optics*, II.3: 80–84). The
same applies also for judging the distance that separates a vertical rod **r** in S_1 from
the observer in S_2, which cannot be determined accurately.

As Ibn al-Haytham argued, the relation with the common ground that is shared
between the observer and the object of vision is measured through the spatiality
of the body of the observer. The feet (*al-qadamayn*) in pacing, the stretched
forearm (*dhira'*) and the hand (*yad*) in grasping, as well as the scale of the human
embodiment (*al-qama*) all act as measure determinants in a pre-reflexive and non-
intentional manner (*Optics*, II.3: 150–155). Ultimately, the estimation of distance
in seeing spatial depth was not restricted to topics in optics, rather they had
applications that were also significant in terms of Ibn al-Haytham's explication of
his observational data in astronomy, like his treatment of the question concerning
the moon-illusion; namely when the moon appears larger at the horizon than at its
zenith.

Ibn al-Haytham's geometrization of place, and his affirmation of the visibility
of spatial depth, resonated with Renaissance and Early-Modern conceptions of
spatiality and extendedness. The definition of place as "space" corresponded also
with the manner architecture and perspective shared a sense of coherent spatiality

as embodied in the "idealized representation" of the notions of the "*room*" and of "a looking space" (Vesely 2004), which acquired the characteristics of the "isotropic space of geometry".[17]

References

Ackerman, J. S. (1949). 'Ars Sine Scientia Nihil Est': Gothic theory of architecture at the Cathedral of Milan. *Art Bulletin, 31*(2), 84–111.

Al-Farisi, K. (1928–1929). *Kitab tanqih al-manazir*, 2 vols. Hyderabad: Osmania Press.

Al-Kindi. (1950–53). *Rasa'il al-Kindi al-falsafiyya*, edited by Muhammad 'Abd al-Hadi Abu Rida (vol. II). Cairo: Dar al-fikr al-arabi.

Al-Kindi. (1997). *Kitab fi 'ilal ikhtilaf al-manazir (De Aspectibus)*. In *Œuvres philosophiques et scientifiques d'al-Kindi*, ed. from Latin and trans. by R. Rashed, Vol. 1: L'optique et la catoptrique. Leiden: E. J. Brill.

Aristotle. (1936). In W. D. Ross (Ed.), *Physics*. Oxford: Oxford University Press.

Barozzi, J. (1611). *Le Due Regole della Prospettiva*. Roma: Camerale.

Carman, C. (2007). Albert and Nicholas of Cusa: perspective as coincidence of opposites. *Explorations in Renaissance Culture, 33*, 196–219.

Crombie, A. (1953). *Robert Grosseteste and the origins of experimental science, 6*. Oxford: Clarendon Press.

Cusanus, N. (1514). *De coniecturis*, in *Opera omnia* (Vol. 1). Paris.

Cusanus, N. (1972). In J. Koch, C. Bormann & I. G. Senger (Eds.), *Nicolai de Cusa Opera Omnia*, Vol. 3. Hamburg: Felix Meiner.

Desargues, G. (1647). *Manière universelle de Monsieur Desargues pour pratiquer la perspective par petit-pied comme le géométral*. Paris: Imprimerie de Pierre des Hayes.

El-Bizri, N. (2004). La perception de la profondeur: Ibn al-Haytham, Berkeley et Merleau-Ponty. *Oriens-Occidens: sciences, mathématiques et philosophie de l'antiquité à l'âge classique. Cahiers du Centre d'Histoire des Sciences et des Philosophies Arabes et Mdivales, CNRS, 5*, 171–184.

El-Bizri, N. (2005a). Ibn al-Haytham. In T. F. Glick, S. J. Livesey & F. Wallis (Eds.), *Medieval science, technology, and medicine: an encyclopedia* (pp. 237–40). London: Routledge.

El-Bizri, N. (2005b). A philosophical perspective on ibn al-haytham's optics. *Arabic Sciences and Philosophy, 15*, 189–218.

El-Bizri, N. (2007a). Imagination and architectural representations. In M. Frascari, J. Hale & B. Starkey (Eds.), *From models to drawings: Imagination and representation in architecture* (pp. 34–42). London: Routledge.

El-Bizri, N. (2007b). In defence of the sovereignty of philosophy: al-baghdadi's critique of ibn al-haytham's geometrisation of place. *Arabic Sciences and Philosophy, 17*, 57–80.

El-Bizri, N. (2009). Ibn al-Haytham et le problme de la couleur. *Oriens-Occidens: sciences, mathématiques et philosophie de l'antiquité à l'âge classique. Cahiers du Centre d'Histoire des Sciences et des Philosophies Arabes et Médiévales, CNRS, 7*, 201–226.

---

[17]This development was perhaps "anticipated" in the "perspectivity" of architecture with the "parallelism" of its structuring components (columns, pillars, walls) and the "axial regularity" of its spatial articulations (Vesely 2004; El-Bizri 2010b).

El-Bizri, N. (2010a). Creative inspirations or intellectual impasses? Reflections on relationships between architecture and the humanities. In S. Bandyopadhyay, J. Lomholt, N. Temple & R. Tobe (Eds.), *The humanities in architectural design: a contemporary and historical perspective* (pp. 123–135). London: Routledge.

El-Bizri, N. (2010b). Classical optics and the perspectiva traditions leading to the renaissance. In C. Carman & J. Hendrix (Eds.), *Renaissance theories of vision* (pp. 11–30). Aldershot: Ashgate.

Euclid. (1883–1916). In J. L. Heiberg & H. Menge (Eds.), *Euclides opera omnia*. Leipzig: Teubner Classical Library.

Euclid. (1956). *The thirteen books of Euclid's elements* (Vols. 1–3) (T. L. Heath, Trans.). New York: Dover Publications.

Federici Vescovini, G. (1965). *Studi sulla prospettiva medievale*, Pubblicazioni della facolta di lettere e filosofia. Turin: Universit di Torino.

Federici Vescovini, G. (1990). La fortune de l'Optique d'Ibn al-Haytham: le livre De aspectibus (Kitab al-Manazir) dans le Moyen Age latin. *Archives d'histoire des sciences*, *40*, 220–238.

Federici Vescovini, G. (1998). Ibn al-Haytham vulgarisé. Le De li aspecti d'un manuscrit du Vatican (moitié du XIVe siècle) et le troisième commentaire sur l'optique de Lorenzo Ghiberti. *Arabic Sciences and Philosophy*, *8*, 67–96.

Federici Vescovini, G. (2008). La nozione di oggetto secondo la Perspectiva di Teodorico di Freiberg. In G. F. Vescovini & O. Rignani (Eds.), *Oggetto e spazio. Fenomenologia dell'oggetto, forma e cosa dai secoli XIII-XIV ai post-cartesiani. Micrologus*, *24*, 81–89 (Firenze: SISMEL, Edizioni del Galluzzo).

Hasse, D. N. (2000). *Avicenna's de anima in the latin west: the formation of a peripatetic philosophy of the soul, 1160–1300*. London–Turin: The Warburg Institute – Nino Aragno Editore.

Ibn al-Haytham. (1983). In A. I. Sabra (Ed.), *Kitab al-Manazir*, 2 vols. Kuwait: National Council for Culture, Arts and Letters.

Ibn al-Haytham. (1989). *The optics, Books I-III, on direct vision* (A. I. Sabra, Trans.), 2 vols. London: The Warburg Institute.

Kemp, M. (1990). *The science of art: optical themes in western art from Brunelleschi to Seurat*. New Haven: Yale University Press.

Koyré, A. (1948). Du monde de l'à peu près à l'univers de la précision. *Critique*, *128*, 806–823.

Lejeune, A. (1948). *Euclide et Ptolémée, deux stades de l'optique géométrique grecque*. Leuven: Bibliothéque de l'Université de Louvain.

Merleau-Ponty, M. (1945). *Phénoménologie de la perception*. Paris: Gallimard.

Nazif, M. (1942–43). *al-Hasan bin al-Haytham, buhuthahu wa-kushufahu al-basariyya*, 2 vols. Cairo: Matba'at al-nuri.

Omar, S. B. (1977). *Ibn al-Haytham's optics: A study of the origins of experimental science*. Minneapolis: Bibliotheca Islamica.

Pelacani da Parma, B. (2002). In G. F. Vescovini (Ed.), *Quaestiones perspectivae*. Paris: J. Vrin.

Rashed, R. (1992). *Optique et mathématiques: Recherches sur l'histoire de la pensée scientifique en arabe*. Aldershot: Variorum.

Rashed, R. (1993). La philosophie mathématique d'Ibn al-Haytham, II: Les Connus. *Les Cahiers du MIDEO*, *21*, 87–275.

Rashed, R. (2002). *Les mathématiques infinitésimales*, Vol. 4. Wimbledon: Al-Furqan Islamic Heritage Foundation.

Rashed, R. (2005). *Geometry and dioptrics in classical Islam*. Wimbledon: Al-Furqan Islamic Heritage Foundation.

Raynaud, D. (2003). Ibn al-Haytham sur la vision binoculaire, un précurseur de l'optique physiologique. *Arabic Sciences and Philosophy*, *13*, 79–99.

Raynaud, D. (2004). Une application méconnue des principes de la vision binoculaire: Ibn al-Haytham et les peintres du trecento (1295–1450). *Oriens-Occidens: Sciences, mathématiques et philosophie de l'Antiquité à l'Âge Classique. Cahiers du Centre d'Histoire des Sciences et des Philosophies Arabes et Médiévales*, *5*, 93–131.

Raynaud, D. (2007). Le tracé continu des sections coniques à la Renaissance: Applications optico-perspectives, héritage de la tradition mathématique arabe. *Arabic Sciences and Philosophy, 17,* 299–345.

Rose, P. (1970). Renaissance Italian methods of drawing the ellipse and related Curves. *Physis, 12,* 371–404.

Sabra, A. (1966). Ibn al-Haytham's criticisms of Ptolemy's *Optics. Journal of the History of Philosophy, 4,* 145–149.

Sabra, A. (1978). Sensation and inference in Ibn al-Haytham's Theory of visual perception. In P. K. Machamer & R. G. Turnbull (Eds.), *Studies in perception: Interrelations in the history of philosophy and science* (pp. 160–185). Columbus: Ohio State University Press.

Sabra, A. (1989). Form in Ibn al-Haytham's theory of Vision. In F. Sezgin (Ed.), *Zeitschrift für Geschichte der Arabisch-Islamischen Wissenschaften* (Vol. 5, pp. 115–140). Frankfurt am Main: Institut für Geschichte der Arabisch-Islamischen Wissenschaften.

Schramm, M. (1963). *Ibn al-Haythams Weg zur Physik.* Wiesbaden: Franz Steiner Verlag.

Simon, G. (1997). La psychologie de la vision chez Ptolémée et Ibn al-Haytham. In A. Hasnaoui, A. Elamrani-Jamal & M. Aouad (Eds.), *Perspectives arabes et médiévales: sur la tradition scientifique et philosophique grecque* (pp. 189–207). Leuven-Paris: Peeters-Institut du Monde Arabe.

Vesely, D. (2004). *Architecture in the age of divided representation: the question of creativity in the shadow of production.* Cambridge, MA: MIT Press.

Chapter 3
The Role of Perspective in the Transformation of European Culture

Dalibor Vesely

The origins of the Renaissance pictorial perspective are closely linked with the transformation of European culture that began already before the fifteenth century. This transformation can be traced back to the new appropriation of nature in the twelfth and thirteenth centuries, growing individualism, the first signs of a new humanism and the change in the nature of knowledge, marked by the return to Aristotelian tradition. However the most important source of perspectival thinking was the new development in the medieval philosophy of light and geometrical optics known then as *perspectiva naturalis*. The move towards a geometrical representation of light was a logical consequence of an attempt to find a more direct form of participation in the essential reality of the divine, closely associated already in the twelfth century with mathematics and in particular with geometry (Ohly 1982, p. 142).

In the treatises of the thirteenth century perspectivists,[1] the properties of light, and not only the physical but also the metaphysical and theological, are discussed almost exclusively in the mathematical language of optics. Apart from the study of vision, optics was used to solve astronomical problems and was used also as a model for a more precise understanding of the nature and the structure of the universe in its totality.

---

[1]The term "perspectiva" was associated with the medieval mathematical optics (*perspectiva naturalis*). Roger Bacon introduced the term as a title of Part V of his *Opus Maius*, originating thus the tradition of *perspectiva* in the West. Bacon had a strong influence on his fellow Franciscan, John Peckham, later Archbishop of Canterbury, who wrote a most popular treatise, *Perspectiva Communis* (1279), and the Silesian scholar Witelo and his long and equally influential treatise *Perspectiva* (1273). The perspectivist tradition persisted through to the fourteenth century. The most influential treatise of that period was Biagio da Parma (Pelacani) *Questiones Perspectivae* (1389).

D. Vesely (✉)
Charles University, Prague, Czech Republic
e-mail: dalibor.v@btinternet.com

R. Lupacchini and A. Angelini (eds.), *The Art of Science*,
DOI 10.1007/978-3-319-02111-9_3,

Fig. 3.1 Brunelleschi: *St Lorenzo*, interior

3.1 The Origins of Perspective

The continuity between medieval optics (*pespectiva naturalis*) and renaissance perspective (*perspectiva artificialis*) played a decisive role in the work of Filippo Brunelleschi, who rightly or wrongly is considered to be the first true renaissance perspectivist. There is no doubt that his work, particularly in architecture, represents a most radical deviation from the late medieval tradition (Fig. 3.1). The internal organisation of his buildings shows entirely new optical unity of space, precisely defined architectural elements, emphasis on the visible manifestation of proportions and what is most radical, lack (negation) of paintings and colours on walls. The walls, particularly in sacred buildings, are left white. The white surface represents a transcendental light as a background for the primary elements of the buildings (columns, arches, architraves etc.) made of darker stone (*pietra serena*) that stand in a strong contrast to the white surface. The churches of St. Lorenzo and St. Spirito are particularly good examples of a new design based on the projection of the spatial depth on the two-dimensional surface. This brought architecture and painting close together and revealed their common ground in pictorial perspective.

The formation of pictorial perspective, sometimes also referred to as *costruzione legittima*, is rightly considered to be the main characteristics of the new historical era and a true revolution in the sphere of visual representation. However the novelty of the new type of representation should not obscure the fact that there is a deep continuity between late medieval and early Renaissance perspective. The decisive step in the new development, which is now accepted as a more or less undisputed initiating event, are Brunelleschi's well-known pictorial demonstrations of the new perspectival method in front of the Florentine Baptistery and in front of the Palazzo della Signoria. There is no need to repeat the details of Brunelleschi's contribution,[2] and what has been achieved in the numerous reconstructions of his experimental demonstration.[3] What is more important and not fully appreciated is the ontological nature and cultural meaning of perspective. There is a tendency, still alive today, to see the formation of pictorial perspective as a problem of correct geometrical representation of vision.[4] However in view of the complex conditions under which it is possible to speak about the true meaning of light and vision geometrical representation itself is not adequate. Genuine representation of light and vision is always situated in an ontological structure of reality (world), which may not be apparent or explicitly visible, but is always assumed. Geometrical construction has in itself no empirical content. If the ontological structure of geometrical operations is not taken into account, the representation as a result remains empty and meaningless.[5]

The development of optics (*perspectiva naturalis*) made it possible to determine visual operations not only as mathematically demonstrable but also in terms of intuitive evidence. What was to be demonstrated in pictorial perspective was the "correctness of sight". But what is the "correctness of sight"? If we take seriously the intentions and contributions of those who took part in the development of pictorial perspective during the fifteenth century it is clear that the correctness of sight cannot be reduced to the correctness of optical structures of representation. The meaning of correctness as it was established at the beginning of the fifteenth century was judged by the degree to which the perfect (divine) order was manifested in representations of the visible world.

The intentions which brought pictorial perspective to existence can be seen as a culmination, and to some extent as a fulfilment, of a development which began at the time of a general orientation of the late medieval culture towards a new appreciation of natural phenomena and the visible world. It is only natural to expect that the privileged position given to vision found its fulfilment in the visual arts and most obviously in painting. However in view of the conventional interpretations and understanding of pictorial perspective it is not easy to see how optics as a mathematical discipline, cultivated in the domain of theology, cosmology, metaphysics, and physics could become the foundation of a new, empirically based, mode of representation. I believe that the key to a more satisfactory understanding of the continuity between medieval optics and Renaissance perspective lies in the

---

[2]For details see Manetti (1927; 1970, pp. 42–46), Filarete (1965, p. 305), Vasari (1965, p. 136).

[3]Important recent contributions to the debate can be found in Klein (1979, pp. 129–143), Parronchi (1964), White (1987, pp. 113–121), Beltrani (1974, pp. 417–468), Edgerton (1975, pp.143–153).

[4]The few exceptions are the contributions of A. Parronchi, S.Y. Edgerton, G. Federici Vescovini, and, to some extent, J. White.

[5]There is a fundamental difference between the role of geometry in the medieval or Renaissance science and in modern science, where it ceases to be part of dialectical reasoning and becomes a pure tool (instrument) of experimental research (Lachterman 1989).

profound change in the representation of reality as a whole, including not only architecture and visual arts, but also everyday life. This change became fully explicit at the beginning of the fifteenth century.

The nature of the change can be characterised as a tendency to represent the traditional, hierarchically structured world as directly accessible and object like. This tendency has in the past been identified with Renaissance individualism and naturalism.[6] However I believe that there is a deeper motivation for the change in a strong desire to recognise the presence of light, intelligibility and order, i.e., the main characteristics of the divine reality, in the human world and to make it accessible through the finite possibilities of human understanding. This may also explain the apparent contradiction in the character of the visual art of the early Renaissance, its illusionistic realism combined with the abstract mathematical rigour of proportional harmonies and perspectival constructions. Proportional reasoning can be seen as a mediating and harmonising link between illusionistic realism and mathematical rigour of perspectival constructions, and in the arguments of some authors (Wittkower, Parronchi, and several others), as the very essence of artificial perspective. The reasoning of proportion follows the articulation of light represented already in medieval optics in the following way (Fig. 3.2):

> It is clear that light through the infinite multiplication of itself extends matter into finite dimensions that are smaller and larger according to certain proportions that they have to one another and thus light proceeds according to numerical and non-numerical proportion. (Grosseteste 1974, p. 12)

Once we leave behind the conventional understanding of proportion as a visible, quantifiable relation between clearly defined entities we discover that proportion is more universal and that it is primarily a qualitative relation. In the non-dogmatic tradition of thinking, proportion is, as the original Greek term "analogia" indicates, an analogy. Analogy is a symbolic structure reflecting the resemblances, similarities and eventually the balanced tension of sameness and difference between individual phenomena. Seen in that light, proportion is a key to the analytical, qualitative articulation of reality and its representation.

The close link between proportion and perspective has been mentioned and emphasised many times. In fact some authors go so far as to believe that the problem of proportionality is the very foundation of perspective, in other words that proportion is "a mathematical concept on which Renaissance theory of perspective rests" (Wittkower 1953). We may agree, but if we do, we have to answer a fundamental question—how is a world structured by analogical proportions and medieval optics represented in the geometrical construction of perspective, which does not seem to express any empirical content, and is, in accord with conventionally understood intentions, a purely formal and universal mathematical discipline—a

---

[6]The association of Renaissance with individualism and naturalism goes back to Jakob Burckhardt (1860) and dominates art-historical writing even today. The problem is discussed in a new and revealing way in Taylor (1989) and Summers (1987).

Fig. 3.2 Cesare Cesariano:
Vitruvius De Architectura,
(Como, 1521). The
multiplication of celestial
light

symbolic "form"?[7] It is taken for granted that the term "symbolic" refers to a
representation of space. However is it not the second term, namely "form", which
tells us that the representation does not refer to the space of our everyday existence
but only to its formal structure? In this case the question of the world, how is it
represented or if it is represented at all, is even more relevant.

In the still ongoing discussion about the nature of artificial perspective it is
not clear if perspective is a symbolic form, i.e., a scientific mathematical mode of
representation, or a rationalisation of concrete visual experience. Almost everyone
seems to agree that the workshop tradition of practical perspective, the medieval
optics as well as the inventiveness of certain artists such as Brunelleschi, Donatello,
Masaccio, Paolo Uccello, and Leon Battista Alberti all played an important role
but it is not yet clear what brought the individual contributions together in the
decisive period when the *costruzione legittima* was formed. Was it the geometry
of the visual pyramid and its projection, the discovery of the vanishing point
or the proportional construction of the foreshortenings? I do not think that the
analytical and technical steps themselves can explain the synthetic nature of the
new perspective. The discovery of the "legitimate construction" was a culminating
point in a long development in which were reflected not only important changes in
the nature of visual arts, but also fundamental transformations in European culture
as a whole. The terms most often associated with this transformation are *devotio*

---

[7]See Cassirer (1923–29) and Panofsky (1924–25). For the critical assessment of the concept
"Symbolic Form", see Boehm (1969), and, as to the Davos Disputation between Ernst Cassirer
and Martin Heidegger, see Heidegger (1929; 1989, pp. 264–268).

moderna in religious life, *via moderna* in the intellectual life and *ars nova* in the domain of arts.[8]

For the purpose of my argument I am interested only in one aspect of the change—the tendency to move away from the hierarchically structured traditional world towards a world in which the transcendental, intelligible levels of reality are seen as immanent and directly visible. The notion which plays the most important role in this change, particularly in the visual arts, was "common sense", the unifying faculty of all senses, the lower unity of meaning and place of "sensible" judgement. The unity of common sense corresponds to the unity of things sensed in terms of their essential characteristics—common sensibles.[9] Typical common sensibles are movement, rest, shape, unity, number, and magnitude which includes sizes and distances.[10] The possibility of seeing magnitudes does not mean that we "can apprehend the exact dimensions or distances of things but that what we apprehend is measurable and corresponds to the measurable" (Summers 1987, p. 153). It is for this reason that the history of common sense is closely bound up with optics.

> Optics in fact might be described as the science of the common sense par excellence, and provides a clear example of the relation between common sense and reason. We always perceive particular shapes and magnitudes under real circumstances and therefore in a certain sense perceive them "incorrectly" and optics tells us what we "really see". (*Ibid.*, p. 83)

The proximity of the judgement of sense and the geometry of vision makes it possible to discern a new relationship between the principles of medieval optics and the achievements of practical workshop kind of perspective already at the end of the fourteenth century. A decisive contribution was the new interpretations and commentaries on medieval optics. The most interesting, from our point of view, are the commentaries of Biagio di Parma (known as Pelacani) and in particular his unpublished treatise *Questiones Perspectivae*.[11] In his writings Biagio, who belongs to the late medieval tradition but also to the epoch of Brunelleschi, though only indirectly, discussed perspective and the questions of vision in a language focused on the tangible visual qualities, on the primary role of common sensibles and on common sense. In his *Questiones Perspectivae*, Biagio is mostly concerned with the question of the judgment of sight (*iudicium sensus*).[12] Such a question can be discussed but cannot be fully answered by verbal argument. For Biagio, the power

---

[8]For the movement "Devotio Moderna", see Post (1968) and Heer (1953).

[9]See Aristotle, *De Anima*, 425b. There is a close affinity between "common sensibles" and Heidegger's categorial intuition (Heidegger 1925).

[10]Leonardo defines a similar common sensibles for painting. "Painting is concerned with all the ten attributes of sight which are—darkness, light, solidity and colour, form and position, distance and propinquity, motion and rest" (Richter 1970, p. 19).

[11]For more details see Federici Vescovini (1980). See also the next footnote.

[12]Biagio's *Questiones Perspectivae* may have been known in Florence already before the end of the fourteenth century but the text was certainly available after Paolo Toscanelli's return to Florence in 1424. See Graziella Federici Vescovini (1960, 1980) and Eugenio Garin (1967).

to decide (*virtus distinctiva*) did not reside in the intellect or in words but in the sight itself. It is in this domain, the domain of visual experience, that the question will be addressed by the next generation.

3.2 The Transformation of the Visible World

The new relationship between the reality articulated by optics and later by linear perspective illustrates the transformation taking place in Florence in the first decade of the fifteenth century. Owing to a unique combination of historical circumstances it was there that it became possible to demonstrate the continuity between the optical interpretation of the medieval world (structured by cosmology and the problems of creation) and the perspectival representation of the directly visible world. We have anticipated this problem in our earlier discussion of the role of geometry in medieval optics, but more needs to be said, particularly concerning the assumptions on which the most decisive steps in the development of linear perspective, including the contribution of Brunelleschi and his *costruzione legittima*, were based.

Pictorial (artificial) perspective was never supposed to be a purely mathematical or absolute discipline, but a pictorial one, representing not a concept of space or abstract structure, but a concrete world in its visibility. In such a world space is not only articulated but it is also embodied and situated, which means that it always has a situational structure as a background to all possible transformations (Merleau-Ponty 1945; 1962, p. 254). The development of perspectival representation was closely linked not only with medieval optics, new treatments of proportions, the imaginary or ideal structure of design (*lineamentum*) but also with surveying, geography and, most of all, with the development of the pictorial space in artists' workshops. The practice of the workshops is particularly important because it was there that the synthetic creative steps occurred.[13]

The first signs of the change towards a new type of pictorial space can be seen in the works of Giotto, his older and younger contemporaries (Cavallini, Cimabue, and Duccio), and his disciples (Taddeo Gaddi). The change in the interpretation of space is always a result of a more fundamental change in the intellectual life and in the sensibility of a particular epoch. The nature of the change cannot therefore be understood in isolation or as a formal problem. The *Presentation of the Virgin* by Taddeo Gaddi in Santa Croce in Florence is a good illustration of such a change in the period of transition from medieval to a proper Renaissance representation. The composition of the painting, dominated by an oblique construction of a temple, is treated in a medieval manner, as a configuration of individual scenes and places in relation to their meaning and not in terms of a unifying space. This is

---

[13]I am using the term "workshop" (*bottega*) in the broadest sense, as a place of work which includes not only the studio type of workshops but also a building site and the working spaces of large commissions.

apparent in the lack of a clear relation between figures and their surrounding, or between themselves, and there is no unity of event, time, and place. The question of realistic unity or unifying space remains problematic in a world structured in accordance with symbolic topology, where imaginative space is important, though the descriptive one is not.

The growing emphasis on the concrete representation of directly visible reality began with the interest in the corporeality of the human body with all its typical characteristics, such as modelling and volume, incidental light, and shadow defining simultaneously body and space. The new interest in a more precise definition of corporeality led also to a new, almost mathematically clear relation between body, its surface and space. Mathematical clarity is manifested most clearly in the geometry of the depicted architectural structures (*cassamenti*). As a paradigm of embodiment and spatiality, architecture became a prime, dominating element in the formation of the new pictorial space and in the process of "perspectivisation". What gave architecture such a privileged position was its idealised, quasi-mathematical nature, the main characteristics of perspective itself.

Architecture and perspective share the same sense of coherent space, most clearly exemplified in the concept of a "room". The space of a room is obviously not the same as the phenomenal space of the natural world. It is a highly idealised representation which during its long history acquired many of the characteristics of the isotropic space of geometry. The natural perspectivity of architecture is already anticipated in the prevailing parallelism of columns, pillars, and walls, in the axiality and in the overall regularity of its spatial arrangement. It is true that perspective depth can be represented by other non-linear means, such as light, shadow, and colour or by perspectival foreshortening of the figures, but even in such situations the sense of room seems to play a decisive role.

Surveying the fourteenth century paintings it is clear that the transformation of pictorial space was mostly accomplished through depicted architecture. This was relevant not only for a new, more unified organisation of space, but also for a new way of representing the traditional medieval order of reality. In the medieval context the individual elements of architecture were closely linked with particular themes and their content (see Sedlmayr 1959; Bandmann 1978). Their purpose was to situate important events and their protagonists in the broader context of reality and its meaning. In that sense the enclosed room-like space became a place where the traditional vertical relations between celestial and terrestrial, divine and human realities could be represented as a horizontal relation between the nearness of the corporeal world and the remoteness of the new quasi-infinite space.[14] In this light the "discovery" of artificial perspective at the beginning of the fifteenth century is not so much a mathematical or technical problem but rather a deep cultural and ontological question.

---

[14]The problem of infinity was discussed until the seventeenth century as a problem of potential and actual infinity. In the human world, only potential infinity was conceivable. The actualisation of infinity is a modern problem which remains still unresolved (Murdoch 1992). See also Koyré (1971, pp. 29–31).

To summarise and repeat, one of the main preconditions for the "discovery" of the "legitimate construction" was the radical transformation of the late medieval culture and in particular the possibility to bring into explicit visibility the highly articulated inherited world, and to reconcile its reality—expressed very often in the language of mathematics—with the particular, concrete phenomena of the finite human reality. It is against this background that Brunelleschi's experiments become more comprehensible. The experiments were motivated by the vision of a new coherent space with a structure derived from the geometry of the visual pyramid, brought into correlation with the perspectival organisation of the directly visible world. The perspectival organisation is not itself directly visible, because it is not an intrinsic characteristic of the visible world, as is very often assumed.

3.3   The Nature of Perspectival Vision

In phenomenal experience we do not see parallel lines as convergent or as a ready-made geometrical projection on the retina. The distance and apparent size of things are not determined by a perspectival view but by the phenomenal structure of the world to which we belong and through which we move in an essentially non-perspectival manner.

> When we look at a road which sweeps before us towards the horizon, we must not say either that the sides of the road are given to us as convergent or that they are given to us as parallel; they are parallel in depth. The perspective appearance is not posited, but neither is the parallelism. (Merleau-Ponty 1945; 1962, p. 261)

Both are products of the conceptual transformation of the original experience. This is clearly expressed in the notion of the judgement of sense (*iudicium sensus*)—used so often in the Renaissance treatises as a reference to judgement and not to the spontaneity of vision. The judgement of sense (vision) defines, I believe, the nature of Brunelleschi's experiments. Much has been written already about the technicalities of these experiments and most of it does not require to be repeated.[15] The still undecided question is the intended meaning of the experiments. Was it a discovery, invention or demonstration of the vanishing point, legitimate construction of illusionistic space, the demonstration of the mathematical nature of vision, or the discovery of the "truth" of vision?

If we take into account all the available evidence, it appears that the main intention behind the experiments was the search for truth, leading not to a discovery or invention of truth, but to an experimental demonstration of its presence in the visible world. What was supposed to be demonstrated was the possibility of a new link between visible reality and the ultimate source of divine truth. It is not

---

[15]The problem was discussed by Damish (1987; 1994, pp. 74–88).

surprising that in the period of a developed via *moderna* the link was found in the mathematical treatment of light and proportion.[16]

In Brunelleschi's experiments the visible reality of the baptistery and its surroundings painted on a small panel (*tavoletta*) and reflected in a mirror is from the beginning seen as a picture, showing already in its natural configuration certain perspectival characteristics, such as potential lateral points, horizon, symmetry, and the axiality of the line of vision. Only the anticipation of the results, suggested to some extent by these characteristics, and supported by the knowledge of the basic principles of optics, could guarantee the relative success of the experiment. In contrast to earlier attempts, still partial, Brunelleschi's demonstration was systematic and addressed space as a three-dimensional continuum, determined by the geometry of the visual pyramid and its projection on the surface of the panel and eventually on the mirror. The critical part of the experiment was the reconciliation of the actual setting and its representation, but most of all the anticipated proportional relation between them. This was demonstrated by the proportion between the height of the panel and its distance from the mirror and the same proportion between the real height of the baptistery and its distance from the original viewpoint.[17]

The mediating role of the mirror is particularly instructive. It illustrates the detached reflective nature of perspective, manifested most clearly in the ambiguous nature of the plane (intersection) situated halfway between the potential and the actual surface.[18] The intersection of the visual pyramid is the key to all the main

---

[16]Nicolas Cusanus (1401–1464) in a symbolic representation of truth used mathematics as a vehicle in an interpretation situated halfway between Nominalism and neo-Platonic mysticism (Watts 1982, pp. 68–72, 93–101).

[17]The picture could be seen in the mirror through a hole in the panel. Brunelleschi claims, "that whoever wanted to look at it should place his eye on the reverse side, where the hole was large and while bringing the hole up to his eye with one hand to hold a flat mirror with the other hand in such a way that the painting would be reflected in it. The mirror was extended by the other hand a distance that more or less approximated in small *braccia* the distance in regular *braccia* from the place he appears to have been when he painted it up to the church of San Giovanni" (Manetti 1927; 1970, p. 44).

For the critical assessment of the inconclusive Brunelleschi's experiments see Parronchi (1958, 1959), Sanpaolesi (1962, pp. 41–53), and R. Klein (1979).

[18]In his *Compendium on the Soul*, Avicenna describes sight as a formation of images in a mirror, and sees thus the eye as a mirror. "The eye is like a mirror and the visible object is like the thing reflected in the mirror by the mediation of air or another transparent body; when light falls on the visible object, it projects the image of the object onto the eye. If a mirror should possess a soul, it would see the image that is formed in it" (Lindberg 1976, p. 49).

The role of mirror in the formation of perspective can be also illustrated by Filarete's argument: "If you should desire to portray something in an easier way, take a mirror and hold it in front of the thing you want to do. Look in it and you will see the outlines of the thing more easily. Whatever is closer or further will appear foreshortened to you. Truly I think that Pippo di ser Brunellesco discovered perspective in this way. It was not used by the ancients, for even though their intellects were very subtle and sharp, still they never used or understood perspective. Even though they exersised good judgment in their works, they did not locate things on the plane in this way and with this rules. You can say that it is false, for it shows you a thing that is not. This is true; nevertheless

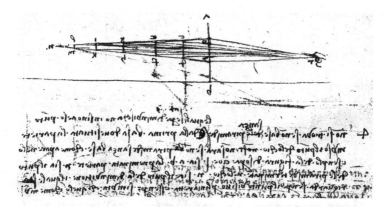

Fig. 3.3 Leonardo da Vinci, manuscript A37r, 1492. Pyramids of vision and perspectival representation

issues of perspective. It is a place where the vanishing point and the horizon are situated and where the pyramid of natural perspective (optics) is reconciled with the visual pyramid in accord with the understanding that "in the practice of perspective the same rules apply to light and to the eye" (Richter 1970, p. 45) (Fig. 3.3).[19]

The structural homogeneity of the two pyramids combined with the empirical identity of the axis of vision constitutes the essence of pictorial representation. In his commentary on the conversion of radiant pyramids emanating from all visible objects into visual pyramids, Leonardo writes:

> perspective is a rational demonstration whereby experience confirms that all objects transmit their similitudes (species) to the eye by a pyramid of lines. (Lindberg 1976, p. 159)

How the similitudes of objects are transmitted by the pyramids of lines we learn from the following, more detailed description.

> Perspective in dealing with distances, makes use of two opposite pyramids, one of which has its apex in the eye and the base as distant as the horizon. The other has the base towards the eye and the apex on the horizon. Now the first includes the visible universe, embracing all the mass of the objects that lie in front of the eye; as it might be a vast landscape seen through a very small opening... The second pyramid is extended to a spot which is smaller in proportion as it is further from the eye; and this second perspective (pyramid) results from the first. (Richter 1970, p. 56)

The experimental demonstration of the legitimate construction in which Brunelleschi, to the best of our knowledge, played the most important role, was fully articulated by Leon Battista Alberti in his treatise on painting (Fig. 3.4). The full articulation meant a completion of the process in which the paradigm of the perspectival room could be reduced to its geometrical essence and fully reconciled

---

it is true in drawing, for drawing itself is not true but a demonstration of the thing you are drawing or what you wish to show" (Filarete 1965, p. 305).

[19]Leonardo's rules appear already in Alhazen's *De Aspectibus* and later in Witelo's *Perspectiva*.

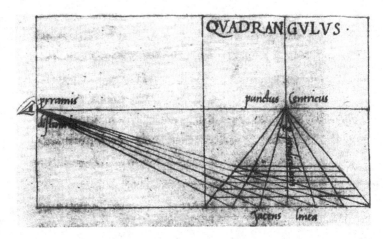

Fig. 3.4 Leon Battista Alberti: *De pictura*, Lucca manuscript, fol. 27r. Perspective diagram

with the geometry of the visual pyramid. Alberti's contribution can be judged only in the light of the results of the experimental period with such impressive achievements as Masaccio's *Trinity* in Santa Maria Novella in Florence (see Dempsey 1972, Polzer 1971, and Goffen 1998) and in the light of the optical knowledge available in his time in Florence.[20]

In terms of substance there was nothing radically new in Alberti's contribution. However in terms of intellectual rigour, conclusiveness and clarity much can be held to his credit. No one before him had the courage to treat the primary issues of perspective as a purely mathematical problem.[21] Alberti's contribution to perspective was developed entirely around the principles of proportion. The sequence of steps that he followed is based on the understanding that the distance and the size of things, projected on the pictorial plane, represent a definite proportion.[22] For the same reason things of the same size situated in the increasing distances from the

---

[20]It is most probable that Alberti became familiar with the primary texts on perspective already during his studies in Bologna (1421–1428). However, it is quite certain that he became familiar with these texts after his return to Florence in 1434. His treatise on painting shows clear indebtedness to John Peckham's *Perspectiva Communis*, Witelo's *Perspectiva*, and probably also to Paolo Toscanelli's treatise on perspective. For the discussion of the text and its possible attribution to Toscanelli, see Parronchi (1964, p. 583).

[21]Alberti's arguments resemble Euclidean geometrical demonstrations leading to axiomatic conclusions. It is rather misleading when he writes in the first book of his *De Pictura* that "I earnestly wish it be born in mind that I speak in these matters not as a mathematician but as a painter" (1436; 1991, p. 37). In truth, he speaks like a mathematician trying to be comprehensible to the painters.

[22]The perspective proportion is based on the well-known proportionality of similar triangles known from Euclidean theorem in Book VI of his *Elements*. It was used in triangulation and in the surveying of tall, distant buildings and objects. Alberti refers to this method in his *Ludi mathematici* (1452).

viewing point appear on the pictorial plane foreshortened in direct proportion to their distances (Wittkower 1953; see also Veltman 1986).

The proportion (*ratio*) of foreshortening can be established in many different ways as we know only too well from the history of perspective. Alberti chose a relatively simple method of sectional projection from a lateral vanishing (distance) point, a method which was sufficiently precise and was not very difficult to understand. The horizontal perspectival grid which can be very quickly obtained by his method is easy to develop into a three-dimensional spatial structure. However it would be misleading to describe this spatial structure as perspective space. At this stage the correct or legitimate construction (*costruzione legittima*) is only a formal representation of space twice removed from reality, first through disembodiment and second as a two-dimensional projection. It is of course true that we are dealing only with the first part of a process which requires to be completed in the re-embodiment of the initial perspectival construction in the pictorial representation of the given visible world. The initial construction may appear as autonomous and mathematically correct, but this cannot be said about its re-embodiment, which can be accomplished only through an imaginative interpretation.

The tension between these two levels of perspective is most often resolved by geometrical formalisation of visual experience and a shift from the epiphanic, essential representation towards a near appearance (illusion) of visual truth. The relativity and the disembodied nature of the new mode of representation can be recognised in the following statement.

> If the sky, the stars, the seas, the mountains and all living creatures, together with all other objects were, the gods willing, reduced to half their size, everything that we see would in no respect appear to be diminished from what it is now. Large, small, long, short, high, low, wide, narrow, light, dark, bright, gloomy and everything of the kind which philosophers termed accidents, because they may or may not be present in things—all these are such as to be known only by comparison, [...] comparison is made with things most immediately known. (Alberti 1436; 1991, p. 45)

Because man is best known to himself, it logically follows that "accidents in all things are duly compared to and known by the accidents to man" (Alberti 1436; 1991, p. 53). This new humanistic position is the foundation of modern relativism as well as of modern aesthetics (Cassirer 1927; Groethuysen 1953; Taylor 1989).

There is an imperceptible sense of power attached to perspective representation which in its capacity to represent mathematically what was believed to be the divine order of reality, made man feel like a god. As Alberti writes "the virtues of painting therefore are, that its masters see their work admired and feel themselves to be almost like the Creator" (Alberti 1436; 1991, p. 6). The conditions under which such feeling could be sustained were defined by the new method of representation and by the precision and overwhelming universality of mathematical method. However it did not take long to discover that perspective is a much more complicated operation than the initial expectations suggested. There is quite clearly a difference between the representation of highly idealised situations used in the early experiments and conceptual models and in the representation of phenomenal reality with a different level of richness and ambiguity.[23]

Alberti himself, as a humanist, was very much aware of the content implied by his method, which he identified with the concept of historia. In fact he goes so far as to say that "the most important part of a painter's work is the historia." In a broader sense, historia is a narrative or programme based on the contribution of narrators and poets (Alberti 1436; 1991, p. 93).

It is typical of Alberti's vision of perspective that he saw the task of representation, including the poetic content, as a quasi-mathematical problem. "Our rudiments," he writes, "from which the complete and perfect art of painting may be drawn, can easily be understood by a geometer, whereas I think that neither the rudiments nor any principles of painting can be understood by those who are ignorant of geometry. Therefore I believe that painters should study the art of geometry" (Alberti 1436; 1991, p. 88). To this end he invented a sequence of steps which made it possible to translate the subtleties of the poetical or rhetorical content into the rigorous language of geometry. Most important was the role of the human body. The content of historia was translated into physiognomic expression, gesture, and movement; the members of the body were structured in final proportions and the composition in accordance with the rules of decorum.[24] The full sequence can be seen as a hermeneutical situation which consists of a relationship between the parts and the whole. "Parts of the historia [...] is the surface, which is defined by lines and angles" (Alberti 1436; 1991, p. 38) and is in that sense a natural element of the geometry of proportion, which means that it can be treated as any other aspect of geometrical perspective.

3.4 From Perspective to *Lineamenta*

How far the represented content of perspective (illusionistic realism) can be expressed by proportional harmonies and perspectival construction is well illustrated in Alberti's discussion of *lineamentum* (Alberti 1485; 1988, p. 7). "The appropriate place, exact numbers, proper scale and graceful order for whole buildings", Alberti claims, can be determined by lines and angles only. In fact he goes one step further when he says:

---

[23] Alberti's own paintings did not survive and there is not even indirect evidence to tell us about their nature, but there is a plausible description of a device he constructed in the form of an optical chamber (*camera ottica*) for the demonstration of perspective construction. "By looking into a box through a little hole one might see great planes and immense expanse of a sea spread out till the eye lost itself in the distance. Learned and unlearned agreed that these images were not like painted things but like nature herself" (Alberti 1843; 1944, p. 284).

[24] Composition was in Alberti's understanding directly linked with the principles of perspective construction. "This method of dividing up the pavement pertains especially to that part of painting which, when we come to it, we shall call composition" (Alberti 1436; 1991, p. 58).

it is quite possible to project whole forms in the mind without any recourse to the material, by designating and determining a fixed orientation and conjunction of the various lines and angles. (*Ibid.*, p. 7)

The imaginary structure of a possible "form" or building anticipates the notion of *disegno interno* (Zuccaro 1607)[25] of the mannerists, but it is still close to the geometrical principles of medieval optics[26] and to the use of geometry in medieval architecture (Simson 1956).[27] In a similar way as *lineamentum*, *disegno interno* belongs to the inventive capacity of the human mind. By means of internal design it is possible to invent an imaginary world as an ideal image, which precedes the realisation of such a world. Zuccaro, the mannerist painter and writer, left a very vivid description of *disegno*:

Man almost imitating God and emulating nature may produce infinite artificial things similar to the natural, and by means of painting and sculpture make us see new paradises on earth.[28]

The similarity between *lineamenta* and medieval geometry shows very clearly the new nature of *lineamenta*; not so much in view of what they represent but how they represent. Unlike medieval geometry, which determines the nature of a particular configuration, such as a portal, a facade, a window, a wall, or interior space—always in view of a unifying whole and in an open dialectical interpretation—*lineamenta* play the same role, but only in respect of the visible unity of the result and as a closed system to which nothing can be added and from which nothing can be taken away.

The possibility of seeing buildings as surfaces defined by *lineamenta* has, no doubt, been prepared by a long history of geometrical interpretation of primary architectural problems (Ohly 1982). This includes the domain of symbolic meanings associated with geometrical optics and its operations, such as the articulation of proportions for instance. It is in the domain of proportions that the difference between *lineamenta* and medieval geometry becomes most visible. In the medieval

---

[25]This issue is extensively discussed by Summers (1987).

[26]Compare with the text of Grosseteste's *De lineis angulis et figures*: "All causes of natural effects must be expressed by means of lines, angles and figures for otherwise it is impossible to grasp their explanation" (Grand 1974, p. 385).

[27]In the late medieval treatise, *Concerning Pinnacle Correctitude*, Mathias Roriczer describes the construction of the pinnacle, which some medieval authors associated with the pyramidal multiplication of light, preserving on each level the similarity (simile) of light to its source (*lux*) (Hedwig 1980, p. 177). In the individual steps of his construction, Roriczer seems to observe the same principle of similarity which he describes as "correct proportion" (*rechtem Mass*): "since each art has its own matter form and measure, I have tried, with the help of God, to make clear this aforesaid art of geometry, and for the first time, to explain the beginning of drawn-out stonework—how and in what measure it arises out of the fundamentals of geometry through manipulation of the dividers, and how it should be brought into the correct proportions" (Shelby 1977, pp. 82–83). There is a close affinity between the proportional sequence in the pyramid of the pinnacle (simile-proportio) and the proportional foreshortening in the perspectival pyramid.

[28]For a more detailed discussion, see Summers (1987, p. 292), Panofsky (1968, pp. 85–93).

context proportions are a direct expression of the hierarchical organisation of space. Adapted as they are to the universe of nature they are the most important means of exploring the secret of a symbolically structured world. In view of all the other options "the only method which can be at all fruitful in such a case is reasoning by analogy and especially the reasoning of proportion" (Gilson 1924, p. 209). In contrast to the symbolic meaning of the medieval geometry, *lineamenta* tend to preserve the same meaning by elevating the imaginative to imaginary level of representation, where the hierarchical organisation of space becomes a coherent, directly visible system of proportions. Alberti's concept of *lineamentum*, expressed most clearly in the already mentioned statement: "it is quite possible to project whole forms in the mind without any recourse to the material" (*ibid.*, p. 7), throws an interesting light on Brunelleschi's treatment of architectural elements in the context of the enclosing space.

Brunelleschi's choice of stone instead of marble, as a material for the primary architectural elements, illustrates his intention to emphasize the neutrality and abstract nature of the elements, avoiding their decorative appearance. This resulted in suppressing the accidental qualities of matter in order to give place to the clearest visibility of the coherent system of proportions. What makes the system coherent is the proportional continuity between individual elements and projective relation between them. This is very clearly demonstrated in San Lorenzo, where the main nave is structured as a precise perspectival projection, completed by the projective relation between the wall of the nave and the wall of the aisle. Here it is not only the transformation of scale, but also the transformation of the three-dimensional elements of the nave into their two-dimensional equivalents on the wall of the aisle, that follows the principles of the perspectival projection.

The novelty and generic role of the primary architectural elements in the work of Brunelleschi was recognised already by his first biographer Antonio Manetti, who describes the elements as "members and bones".[29] The visual and material separation of the elements from the surface of the wall creates a tension between the body of the building and its essential elements, represented by columns, pillars, arches, entablatures, architraves, and frames of the openings. They all appear as clearly defined on the background of the white surface of the walls. Brunelleschi's choice of the neutral white walls can be explained by Alberti's reference to Cicero and Plato, who "reject the variety and frivolity in the ornament of their temples and value purity above all else".[30] There is however also a different, possible explanation. The vanishing point in perspective designates the ultimate depth in relation to infinity. This is not easy to visualise, as is quite clear from the variety

---

[29]"He [Brunelleschi] seemed to recognise very clearly a certain arrangement of members and bones (*il conoscere un certo ordine di membri e d'ossa*) just as if God had enlightened him about great matters" (Manetti 1927; 1970, p. 51).

[30]From this reference Alberti drew his own conclusion: "I would easily believe, that in their choice of colour, as in their way of life, purity and simplicity would be most pleasing to the gods above, nor should a temple contain anything to divert the mind away from religious meditation towards sensual attraction and pleasure" (Alberti 1485; 1988, pp. 219–220).

of solutions in the Quattrocento paintings, that situate most often the vanishing point in some zone (plane) of indeterminacy. The zone of indeterminacy can be seen as a zone of transcendence as a source of transcendental light in a similar way as the white or gold background in the medieval manuscripts. It is in relation to this background that we can appreciate better the meaning of Brunelleschi's primary architectural elements, their relation to Alberti's *lineamentum*, but also to the mannerists *disegno interno*. These relations are supported by the neo-Platonic way of thinking in which the source of light is also the source of intelligibility. The *disegno interno* was for the mannerists a supreme form of intelligibility, since the human intellect by virtue of its participation in God's ideational ability and its similarity to the divine mind as such, can produce in itself the intelligible forms of all created things and can transfer these forms to matter. Zuccaro (II.16. p. 196) interprets the term disegno interno as symbol of man's similarity to God (disegno—segno di dio in noi).

As a universal formative power, *disegno interno* can be seen as a general source of creativity, which can help to grasp the essential nature of sensible phenomena in a new kind of creative process. What is new in this process is the premise that which is to be revealed in a work of art must first be present in the mind of the artist. The introverted presence of what is to be revealed is a quasi-idea, which, as the mannerists believed, man shares with God and makes man thus God-like. The idea (idealised reality) that precedes what is to be discovered (revealed) in reality, transcends the intentions of the mannerists. It became a foundation of modern form of knowledge and as a result of modern European culture as a whole. The link between Brunelleschi's invention of perspectival construction, the *lineamenta* and *disegno interno* became the foundation of modern "scientific" optics.

3.5 From Perspective and Optics to Modern Science

The dialogue between perspectival constructions and phenomenal reality and their influence on the nature of optics underwent a radical change in the sixteenth century. The first signs of this can be found in the works of significant artists of the time such as Leonardo da Vinci, Jan van Eyck, Albrecht Dürer, and Holbein, to mention but a few. They began to use convex mirrors, lenses, and mechanical devises for the construction of perspective and, in case of Leonardo and Holbein, also the *camera obscura* (Huerta 2003, p. 25). The creative nature of perspective had, by this time, changed into a dialogue between the instruments of enforced perspectivity and controlled observations (Fig. 3.5). This represented altogether a shift from a cosmological understanding of light to a purely optical understanding, focused on the nature and truth of vision. The change became a foundation for a new type of optics initiated and legitimized by Johannes Kepler in his seminal text on optics *Ad Vitelonem Paralipomena quibus Astronomiae Pars Optica Traditur (1604)*. As the title already suggests, Kepler's point of departure was not optics but astronomy, nevertheless it was his observation of the eclipse of the moon using a camera

Fig. 3.5 Sebastian Le Clerc: *The academy of sciences and fine arts*, frontispiece, Paris 1666

obscura that raised questions regarding the formation of images which developed into the main concern of the new optics. Produced by light, images, in Kepler's understanding, were mere causal effects carried by light which accidentally reflected objects and fell on a screen. Kepler's optics had no room for forms (species) and visual rays, and without them the fulfillment of the optical process as teleology was lost together with the essential truth of vision and the importance of optics as a source of knowledge for all other sciences; understanding of the "true" vision Kepler leaves, as he declares, to the philosophers.[31]

Kepler's own understanding of optics was based on the assumption that true vision could be grasped by the neutral objectivity of optical instruments because the resulting images—outcomes of purely causal process which we can investigate through experiment and geometrical reasoning—can be trusted. Therefore, for Kepler, the key to the truth (correctness) of vision was to be found in the anatomy and optical behavior of the eye, his knowledge of which was based on the texts

---

[31]"I shall describe the means of vision, which no one at all to my knowledge has yet examined and understood in such detail. I therefore beg the mathematicians to consider this carefully, so that thereby at last there might exist in philosophy something certain concerning this most noble function. I say that vision occurs when an image of the whole hemisphere of the world that is before the eye, and a little more, is set up at the white wall, tinged with red, of the concave surface of the retina. How this image or picture is joined together with the visual spirits [*species*] that reside in the retina and in the nerve, and whether it is arraigned within by the spirits into the caverns of the cerebrum to the tribunal of the soul or of the visual faculty, given by the soul whether the visual faculty, like a magistrate given by the soul, descending from the headquarters of the cerebrum outside to the visual nerve itself and the retina, as to lower courts, might go forth to meet this image—this I say , I leave to the natural philosophers to argue about". (Kepler 1604; 2000, p. 180).

of the experts as he had no direct experience in this regard.[32] Passively receiving illuminations like any instrument, the eye, in Kepler's understanding, is not merely comparable to *camera obscura*; it is one. The pupil has taken the place of Alberti's perspectival window; the cornea is now nothing but a lens and the retina nothing but a screen. Kepler was the first to declare that genuine vision occurs when the pupil of the eye is exposed most closely to the arriving ray of light. In this understanding of vision the human observer has disappeared from the treatment of optics. The naturalisation of the eye, in which the natural and artificial became one, separated experience from its objects. Turned into an optical instrument the eye no longer furnished the observer with genuine representation of visible objects. It became a mere screen on which an anonymous image is projected. In the case of the eye, the screen was identified as retina and the projected image as picture.[33] The description of the projected image as picture is not a metaphor but, at this time, should be taken literal. It is well known that in the Dutch art of the seventeenth century there is a close link between painting and the use of optical instruments—mainly the *camera obscura*—as is evident in the paintings of Vermeer, Hoogstraten, Hooch, and many others.[34] Their works can be seen not only as paintings but also as a confirmation of the new optics and vision of reality in which the artificial and natural, the manmade picture and the visible reality, are one.

Kepler readily acknowledged his indebtedness to the perspectivist tradition, but his optics were no longer an account of how the "visible object" recreated its "likeness" in the eye; it was a mathematical-physical theory of the formation of images by light. This represents a beginning of a tradition that is best summarized by Descartes' interpretation of light and optics, based to a great extant on the

---

[32]"Let men of accepted authority speak for me on the subject that is best known to them, up to the point where the undertaking shall have reverted to the limits of my profession. For then they too will ungrudgingly hand the torch over to me at the point where I am going to carry it forward legitimately into mathematics, concerning which the judgment will belong to the expert. I have consulted chiefly Felix Platter's plates concerning the structure and use of the human body, which, published in 1583 were deservedly reprinted in this year 1603. With these I compared the *Anatomia Pragensis* of my friend Mr. Johannes Jessenius of Jessen, for the reason that he not only professed chiefly to follow Aquapendente but on his own prowess devoted himself chiefly to anatomical labors. If I, being myself chiefly occupied in the mathematical profession, have passed over any of greater merit in the succession, they will grant me pardon". (Kepler 1604; 2000, p. 171)

Felix Platter (1536–1614) was a professor of medicine in Basel and wrote *De partium corporis humani structura et usu libri III* (Basel: 1583 and 1603). Johannes Jessenius a Jessen (1566–1621) was a professor of medicine in Wittenberg and later in Prague where Kepler came to know him. He wrote *Anatomiae Pragae anno 1600 ab se solemniter administrate historia* (Wittenberg: 1601). Hieronymus Fabricius Aquapendente (1537–1619), wrote *De visione, voce, et auditu* (Venice: 1600).

[33]"From the Sun and the colors illuminated by the Sun, species flow until for whatever reason, they fall on an opaque medium, where they paint their source: and vision is produced, when the opaque screen of the eye is painted this way, for there are certain passions of light illuminating and altering the screens [of the eye] through which colors, that is to say light, are not only poured upon but are also imprinted". (Kepler 1604; 2000, pp. 41–42).

[34]See Alpers (1983).

Optics of Kepler. Descartes' theory is the first clearly to assert that light itself was nothing but a mechanical property of the luminous object and of the transmitting medium. His theory was the starting point of modern physical optics, useful for the construction of optical instruments (telescopes, microscopes etc.) but indifferent to the understanding of the content of vision and the differentiated, qualitative content of the visible world. Kepler's *Optics* is not only a turning point, but also a culmination of the development of modern perspective that began with Brunelleschi. The dialogue between the perspectival construction and the visible world in the time of Renaissance was replaced in the time of Kepler by experimental dialogue, in which the anonymous objectivity of the geometrical construction of vision was reconciled with the given visible reality by the use of optical instruments and critical observations. This brought the development of perspective, in a form of new optics to the same level as the modern experimental science, formed for the first time consistently by Galileo's own experimental method.

References

Alberti, L. B. (1436). *De pictura*. English translation: *On painting*. London: Penguin Books (1991).

Alberti, L. B. (1452). *Ludi mathematici*. In C. Grayson (Ed.), *Opere Volgari*. Bari: Laterza (1973).

Alberti, L. B. (1485) *De re aedificatoria*. English translation: *On the art of building*. Cambridge, MA: MIT Press (1988).

Alberti, L. B. (1843). *Vita Anonima*. In Bonucci Anicio (Ed.), *Opere Volgari de Leon Battista Alberti* (Vol. I, pp. cii–civ). *Proceedings of British Academy*, 30, Florence (1944).

Alpers, S. (1983). *The art of describing, Dutch art in the seventeenth century*. Chicago: University of Chicago Press.

Bandmann, G. (1978). *Mittelalterliche Architektur als Bedeutungsträger*. Mann, Berlin: Verlag Gebr.

Beltrani, R. (1974). Gli esperimenti prospettici del Brunelleschi. *Rendiconti della Sedute dell'Academia Nazionale dei Lincei (classe di Scienze morali storiche e filologiche)*, XXVIII, 417–468.

Boehm, G. (1969). *Studien zur Perspektivität*. Heidelberg: C. Winter.

Burckhardt, J. (1860). *Die Kultur der Renaissance in Italien*. English translation: *Civilization of the Renaissance in Italy* (2 vols). London: George Allen (1878).

Cassirer, E. (1927). *The individual and the cosmos in renaissance philosophy*. Philadelphia: University of Pennsylvania Press (1963).

Cassirer, E. (1923–29). *Philosophie der Symbolischen Formen*, Berlin: B. Cassirer.

Damish, H. (1987). *L'origine de la perspective*. Paris: Flammarion [English translation: *The origin of perspective*. Cambridge: MIT Press (1994)].

Dempsey, C. (1972). Masaccio's trinity: Altarpiece or tomb?. *Art Bulletin*, *54*, 279–281.

Edgerton, S. Y. (1975). *The renaissance rediscovery of linear perspective*. New York: Harper & Rowe.

Federici Vescovini, G. (1960). Biagio Pelacani da Parma. *Rivista di Filosofia*, *LI*, 179–185.

Federici Vescovini, G. (1980). La prospettiva del Brunelleschi, Alhazen e Biagio Pelacani a Firenze. In *Filippo Brunelleschi, la sua Opera e il suo Tempo*. Florence: Centro Di.

Filarete (ca 1460). *Libro architettonico*. English translation by J. R. Spencer: *Treatise on architecture*. New Haven: Yale University Press (1965).

Garin, E. (1967). *Ritratti di Umanisti*. Florence: Sansoni.

Gilson, E. (1924). *The philosophy of St. Bonaventura*. Paterson, NY: St. Anthony's Guild Press (1965).

Goffen, R. (Ed.). (1998). *Massaccio's trinity (masterpieces of western painting)*. Cambridge: Cambridge University Press.

Grand, E. (Ed.). (1974). *A source book in medieval science*. Cambridge, MA: Harvard University Press.

Groethuysen, B. (1953). *Anthropologie Philosophique*. Paris: Gallimard.

Grosseteste, R. (ca 1220–1230). *De luce*. English translation: *On light* [In Grand (1974)].

Hedwig, K. (1980). *Sphaera lucis: Studien zur Intelligibilität des Seienden im Kontext der mittelalterlichen Lichtspekulation*. Münster: Aschendorff.

Heer, F. (1953). *Europäische Geistesgeschichte*. English translation by J. Steinberg: *The intellectual history of Europe*. London: Weidenfeld and Nicolson (1966).

Heidegger, M. (1925). *Prolegomena zur Geschichte des Zeitbegriffs*. Frankfurt am Main: Vittorio Klostermann [English translation: *History of the concept of time*. Bloomington: Indiana University Press (1985)].

Heidegger, M. (1929). *Kant und das Problem der Metaphysik*. Bonn: Friedrich Cohen [English translation: *Kant and the problem of metaphysics* (together with a protocol of the Davos Disputation). Evanston: Northwestern University Press (1989)].

Huerta, R. D. (2003). *Giants of Delft: Johannes Vermeer and the natural philosophers, the parallel search for knowledge during the age of discovery*. Lewisburg: Bucknell University Press.

Kepler, J. (1604). *Ad Vitelonem paralipomena quibus astronomiae pars optica traditur*. Frankfurt am Main: C. Marnius & heirs of J. Aubrius [English translation by W. H. Donahue: *Optics: Paralipomena to Witelo and the optical part of astronomy*. Santa Fe, New Mexico: Green Lion Press (2000)].

Klein, R. (1979). Studies on perspective in the renaissance. In *Form and meaning*. New York: The Viking Press.

Koyré, A. (1971). L'infini et le contenu. In *Études d'Histoire de la Pensée Philosophique*. Paris: Gallimard.

Lachterman, D. R. (1989). *The ethics of geometry*. London: Routledge.

Lindberg, D. C. (1976). *Theories of vision from al-Kindi to Kepler*. Chicago: University of Chicago Press.

Manetti, A. (1927). *Vita di Filippo di ser Brunellesco*. Florence: Editio princeps, Rinascimento del libro. [English translation by C. Enggast: *The life of Brunelleschi*. Pennsylvania: State University Press (1970)].

Merleau-Ponty, M. (1945). *Phénoménologie de la Perception*. Paris: Gallimard [English translation: *Phenomenology of perception*. London: Routledge & Kegan Paul (1962)].

Murdoch, J. E. (1992). Infinity and continuity. In *The Cambridge history of later medieval philosophy*. Cambridge: Cambridge University Press.

Ohly, F. (1982). Deus Geometra. In *Interdisciplinare Forschungen zur Gesch. des Fruh. Mittl. (IFGFM)*.

Panofsky, E. (1924–25). Die Perspektive als 'Symbolische Form'. *Vorträge der Bibliothek Warburg*. Leipzig-Berlin: B.G. Teubner (1927) [English translation: *Perspective as symbolic form*. New York: Zone Books (1997)].

Panofsky, E. (1968). *Idea: A concept in art theory*. Columbia: University of South Carolina Press.

Parronchi, A. (1958). Le due tavole prospettiche del Brunelleschi. *Paragone, 107*, 3–31.

Parronchi, A. (1959). Le due tavole prospettiche del Brunelleschi. *Paragone, 109*, 3–31.

Parronchi, A. (1964). *Studi sulla dolce prospettiva*. Milan: Aldo Martello.

Polzer, J. (1971). The anatomy of Massaccio's holy trinity. *Jahrbuch der Berliner Museen, 93*, 18–59.

Post, R. R. (1968). *The modern devotion: Confrontation with reformation and humanism*. Leiden: Brill.

Richter, J. P. (1888). *The notebooks of Leonardo da Vinci* (Vol. I). Compiled and Edited from the Original Manuscripts. New York: Dover Publications (1970).

Sanpaolesi, P. (1962). *Brunelleschi*. Florence: G. Barbèra.

Sedlmayr, H. (1959). Über eine mittelalterliche Art des Abbildens. In *Epochen und Werke* (2 vols.) Wien: Herold.

Shelby, L. R. (1977). *Gothic design techniques: The fifteenth-century design booklets of mathes Roriczer and Hans Schmuttermayer*. Carbondale: Southern Illinois University Press.

Simson, O. von (1956). Measure and light. In *The gothic cathedral* (Vol. 48). Princeton: Bollingen Series.

Summers, D. (1987). *The judgement of sense, renaissance naturalism and the rise of aesthetics*. Cambridge: Cambridge University Press.

Taylor, C. H. (1989). *The sources of the self*. Cambridge: Cambridge University Press.

Vasari, G. (1550). *Le vite de' più eccellenti pittori, scultori, e architettori*. Florence: Torrentino [English translation: *The lives of the artists*. Harmondsworth: Penguin Books (1965)].

Veltman, K. H. (1986). Proportionality and perspective. In *Linear perspective and the visual dimensions of science and art*. Munchen: Deutscher Kunstverlag.

Watts, P. M. (1982). *Nicolaus cusanus, a fifteenth century vision of man*. Leiden: Brill.

White, J. (1957). *The birth and rebirth of pictorial space*. London: Faber & Faber.

Wittkower, R. (1953). Brunelleschi and proportion in perspective. *JWCI, 16*, 275–291.

Zuccaro, F. (1607). *L'idea de' pittori, scultori e architetti*. Torino. In D. Heikamp (Ed.), *Scritti d'arte di Federico Zuccaro*. Firenze: Olschki (1941).

Chapter 4
Visual Differential Geometry
and Beltrami's Hyperbolic Plane

Tristan Needham

Historical wrongs are hard to right. In 1868 Eugenio Beltrami (Fig. 4.1) set the previously abstract hyperbolic geometry of Lobachevsky and Bolyai upon a firm and intuitive foundation by interpreting it as the intrinsic geometry of a negatively curved surface. Furthermore, he discovered *all* of the most useful models of this geometry employed today. Yet these models now bear the names of Poincaré and of Klein, while the name Beltrami languishes in semiobscurity.[1]

Hyperbolic Geometry. For more than 2,000 years all mathematicians believed in Euclidean geometry as the correct description of the physical space we inhabit. In particular, the five axioms from which Euclid sought to derive all geometric theorems were themselves understood to be plain facts of Nature.

The fifth and last of Euclid's axioms, dating from 300 BCE, dealt with drawing lines in a plane:

Parallel Axiom. Through any point p not on the line L there exists precisely one line that does not meet L.

But the character of this axiom was more complex and less immediate than that of the first four, and mathematicians began a long struggle to dispense with it as an

---

[1]To add insult to injury, the average mathematician also does *not* remember Beltrami as the discoverer of the singular value decomposition of linear algebra! See Stewart (1993). The resurrection of Beltrami's reputation began with Milnor (1982) and was driven forward decisively by Stillwell (1996, 2010), later assisted by others (e.g., Needham 1997; Penrose 2005). The present essay continues that good fight.

T. Needham (✉)
Department of Mathematics, University of San Francisco, San Francisco, CA 94117
e-mail: needham@usfca.edu

R. Lupacchini and A. Angelini (eds.), *The Art of Science*,
DOI 10.1007/978-3-319-02111-9_4,
© Springer International Publishing Switzerland 2014

Fig. 4.1 Eugenio Beltrami (1835–1900)

assumption, instead seeking to show that it must be a logical *consequence* of the first four axioms.

Many attempts were made to prove the parallel axiom, and the number and intensity of these efforts reached a crescendo in the 1700s, but all met with failure. Yet along the way useful *equivalents* of the axiom emerged. For example: *There exist similar triangles of different sizes* (see Stillwell 2010). But the very first equivalent was already present in Euclid, and it is the one still taught to every school child: *The angles in a triangle add up to two right angles.*

The explanation of these failures only emerged around 1830, when Nikolai Lobachevsky and Janos Bolyai independently announced a revolutionary new non-Euclidean geometry (now called *hyperbolic geometry*) taking place in a new kind of plane (now called the *hyperbolic plane*). In this geometry the first four axioms hold but the parallel axiom does *not*. Instead the following is true:

Hyperbolic Axiom. *Through any point p not on the line L there exist at least two lines that do not meet L.*

$$(4.1)$$

In this strange geometry, it can be shown that the angles in a triangle must add up to *less* than two right angles!

This non-Euclidean geometry had in fact already manifested itself in various branches of mathematics throughout history, but always in disguise. Poincaré was the first not only to strip its camouflage but also to recognize and exploit its power in such diverse subjects as complex analysis, differential equations, number theory, and topology. Its continued vitality and centrality in the mathematics of today is demonstrated by Thurston's work on three-manifolds, and Wiles's proof of Fermat's Last Theorem, to name but two.

But for decades following its initial discovery, hyperbolic geometry was either ignored or else attacked as nonsense. The ultimate acceptance and vindication came only with Beltrami's concrete interpretation, which in turn hinged on concepts of *differential geometry*, the geometry of space that is *curved*.

In an effort to make the discussion as self-contained as possible, the next section traces the very beginnings of differential geometry, reviewing the curvature of plane curves, and explaining the "Newtonian" style of geometric reasoning that we shall employ.[2]

4.1 Newton's "Crookednesse"

Shortly before his Christmas-day birthday[3] in 1664, the 21-year-old Newton began to investigate what he called the "crookednesse" of plane curves (Newton 1967), thereby introducing the concept of curvature into mathematics for the first time.

The *circle of curvature* at a point p on a curve is the one that best approximates the curve in the immediate vicinity of p, just as the tangent is the line that does this best. See Fig. 4.2. Newton constructed the center c (the *center of curvature*) of this approximating circle as the limiting position of the intersection of the normal at p with the normal at a neighboring point q, in the limit $q \to p$. Then pc is called the *radius of curvature*, and $\kappa \equiv (1/pc)$ is what Newton initially dubbed the "crookednesse," but later rechristened as the *curvature*.

As we have discussed elsewhere (Needham 1993; Needham 1997, Preface), Newtonian scholars (see Arnol'd 1990; Bloye and Huggett 2011; de Gandt 1995; Guicciardini 1999; Newton 1999, p. 123; Westfall 1980) have painstakingly dismantled the pernicious myth that the results in the 1687 *Principia* (Newton 1999) were first derived by Newton using his original 1665 version of the calculus, and only later recast into the geometric form that we find in the finished work.

Instead, it is now understood that by the mid-1670s, having studied Apollonius, Pappus, and Huygens in particular, the mature Newton became disenchanted with the form in which he had originally discovered the calculus in his youth—which is different again from the Leibnizian form we all learn in college today—and had instead embraced purely geometric methods.

---

[2]Our book *Visual Complex Analysis* (Needham 1997) arose from the application of this Newtonian approach to complex analysis, and the present essay contains a handful of ideas taken from a sequel, currently in progress, entitled *Visual Differential Geometry*.

[3]In our household we refer to this event as "Newtonmas"!

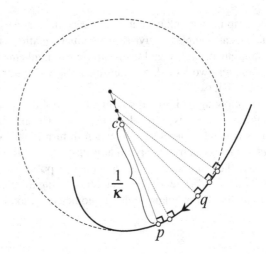

Fig. 4.2

Thus it came to pass that by the 1680s Newton's algebraic infatuation with power series gave way to a new form of calculus—what he called the "synthetic method of fluxions"[4]—in which the geometry of the Ancients was transmogrified and reanimated by its application to shrinking geometric figures in their moment of vanishing. *This* is the potent but non-algorithmic form of calculus that we find in full flower in his great *Principia* of 1687.

Let us spell this out, so that we may take advantage of Newton's approach in the rest of this essay. If two quantities A and B depend on a small quantity ϵ, and their ratio approaches unity as ϵ approaches zero, then we shall avoid the more cumbersome language of limits by following Newton's lead in the *Principia*, saying simply that "A is ultimately equal to B." Also, as we did in an earlier paper (Needham 1993), we shall employ the symbol $\asymp$ to denote this concept of ultimate equality.[5] In short,

$$\text{``A is ultimately equal to B''} \quad \Longleftrightarrow \quad A \asymp B \quad \Longleftrightarrow \quad \lim_{\epsilon \to 0} \frac{A}{B} = 1.$$

It follows from the theorems on limits that ultimate equality is an equivalence relation, and that it also inherits additional properties of ordinary equality, e.g., $X \asymp Y \,\&\, P \asymp Q \Rightarrow X \cdot P \asymp Y \cdot Q$, and $A \asymp B \cdot C \Leftrightarrow (A/B) \asymp C$.

Before we begin to apply this idea in earnest, we also note (again following Newton) that the jurisdiction of ultimate equality can be extended naturally to

---

[4]See (Guicciardini 2009, Chap. 9).

[5]This notation was subsequently adopted by the Nobel physicist Subrahmanyan Chandrasekhar (1995, p. 44).

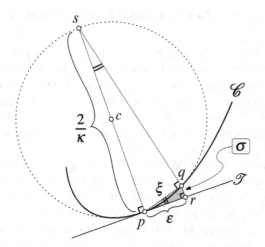

Fig. 4.3

things other than numbers, enabling one to say, for example, that two triangles are "ultimately similar," meaning that their angles are ultimately equal.

Let us immediately make use of this language and notation to derive a result that we shall need in the next section. Figure 4.3 shows a plane curve $\mathscr{C}$ and its circle of curvature at a point p. By definition of the curvature κ at p, the illustrated diameter $ps = (2/\kappa)$. Now let q be a point on $\mathscr{C}$ near to p (where $\xi = pq$) and drop a perpendicular $qr = \sigma$ from q to the tangent $\mathscr{T}$ at p, and finally let $\epsilon = pr$.

Since $\mathscr{T}$ is tangent to $\mathscr{C}$, $\lim_{\epsilon \to 0}(\sigma/\epsilon) = 0$, and therefore

$$\frac{\xi^2}{\epsilon^2} = \frac{\epsilon^2 + \sigma^2}{\epsilon^2} = 1 + \left[\frac{\sigma}{\epsilon}\right]^2 \asymp 1 \implies \xi \asymp \epsilon.$$

Also, the shaded triangle prq is ultimately similar to the triangle sqp, so

$$\frac{\xi}{\left[\frac{2}{\kappa}\right]} \asymp \frac{\sigma}{\xi}.$$

This is essentially Newton's Lemma II, from Book I of the *Principia* (Newton 1999, p. 439) (see also Brackenridge and Nauenberg 2002, p. 112). Depending on whether it is κ or σ that needs to be found in terms of the other, we may combine the previous two results to deduce that

$$\kappa \asymp \frac{2\sigma}{\epsilon^2} \quad \text{or} \quad \sigma \asymp \frac{1}{2}\kappa\epsilon^2. \tag{4.2}$$

4.2   Surface Theory Before Gauss: Euler's Formula

In 1760 Euler analyzed the bending of a surface S at a point p by considering the plane curve $\mathscr{C}_\varphi$ through p obtained by intersecting S with a plane Π_φ that rotates about the surface normal $\mathbf{N}_p$ at p. See Fig. 4.5, which illustrates two orthogonal positions of this plane Π_φ, on two different kinds of surface. Here φ denotes the angle of rotation of Π_φ, starting from an arbitrary (at least for now) initial direction. As Π_φ rotates, the shape of the intersection curve $\mathscr{C}_\varphi$ changes, and therefore its curvature $\kappa(\varphi)$ at p will (in general) vary too.

Before continuing, we should explain that $\kappa(\varphi)$ has a *sign* attached to it, according to this convention: the vector from p to the center of curvature c of $\mathscr{C}_\varphi$ is defined to be $\frac{1}{\kappa(\varphi)}\mathbf{N}$. Thus if c lies in the direction of $+\mathbf{N}$, then $\kappa(\varphi)$ is positive, while if it lies in the direction of $-\mathbf{N}$, then $\kappa(\varphi)$ is negative. Of course there are actually two opposite choices for $\mathbf{N}$, but once we (arbitrarily) choose one of these, we may continuously extend this choice over the whole (orientable) surface.

As φ varies, let κ_1 and κ_2 denote the maximum and minimum values of $\kappa(\varphi)$. Euler's elegant and important discovery was that these extreme values of the curvature [the so-called *principal curvatures*] will always occur in *perpendicular* directions, which are called the *principal directions*. Furthermore, choosing $\varphi = 0$ to coincide with the direction that has curvature κ_1, he found

$$\text{Euler's Formula: } \kappa(\varphi) = \kappa_1 \cos^2 \varphi + \kappa_2 \sin^2 \varphi.$$

Note that the extremal nature of κ_1 and κ_2, together with the orthogonality of the principal directions, can be deduced directly from this formula, as becomes clear when it is rewritten[6] as

$$\kappa(\varphi) = \left[\tfrac{\kappa_1 + \kappa_2}{2}\right] + \left[\tfrac{\kappa_1 - \kappa_2}{2}\right] \cos 2\varphi,$$

the meaning of which is made plain by its graph in Fig. 4.4.

Gauss was the first to realize that the *product* of the principal curvature has deep significance, and in his honor this is called the *Gaussian curvature*, $\mathscr{K} \equiv \kappa_1 \kappa_2$. To begin to see how this quantity governs the shape of the surface, we note that while the signs of the principal curvatures depend on the arbitrary choice of $\mathbf{N}$, the sign of the Gaussian curvature $\mathscr{K} = \kappa_1 \kappa_2$ does not.

As illustrated in Fig. 4.5a, if $\mathscr{K} > 0$ then κ_1 and κ_2 have the same sign, and $\kappa(\varphi)$ always shares this same sign, i.e., the surface locally resembles a hump. But if $\mathscr{K} < 0$, then $\kappa(\varphi)$ changes sign, i.e., $\mathscr{C}_\varphi$ flips from one side of the tangent plane to the other, and, as shown in Fig. 4.5b, the surface locally resembles a saddle.

---

[6]To do this, recall that $\cos^2 \varphi = (1 + \cos 2\varphi)/2$ and $\sin^2 \varphi = (1 - \cos 2\varphi)/2$.

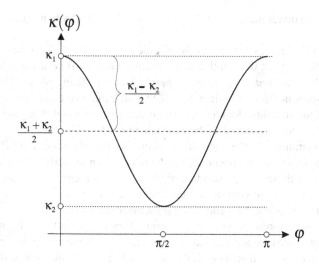

Fig. 4.4

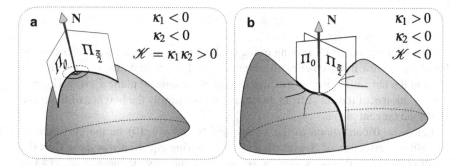

Fig. 4.5

We will now provide a mainly geometric proof[7] of Euler's Formula. Choose p to be the origin of the Cartesian (x, y, z) coordinates, and let the x and y axes be chosen to lie in the tangent plane T_p at p. Then locally the surface can be represented by an equation of the form $z = f(x, y)$, such that $f(0, 0) = 0$ and $\partial_x f = 0 = \partial_y f$ at the origin. Expanding $f(x, y)$ into a Taylor series, we deduce that as x and y tend to zero,

$$z \asymp ax^2 + bxy + cy^2. \tag{4.3}$$

Slicing through the surface with planes $z = \text{const.} = k$ parallel to T_p, and very close to it, therefore yields intersection curves whose equations [as k goes to zero]

---

[7]The proof in the lovely article by A.D. Aleksandrov (1969) is similar, but even that requires two calculations, which are here replaced with geometry.

are (ultimately) quadratics, $ax^2 + bxy + cy^2 = k$, and which are therefore conic sections.

Figure 4.5 illustrates the fact that these conics are ellipses if $\mathcal{K} > 0$ and that they are hyperbolas if $\mathcal{K} < 0$. In both cases, *the conics have two perpendicular axes of symmetry which are independent of the height k of the slicing plane.* This follows from the homogeneous quadratic nature of the equation. For example, quadrupling the height of the slice just doubles the size of the conic, without changing its shape: $k \rightarrow 4k$ yields the same curve as the expansion $(x, y) \rightarrow (2x, 2y)$.

Thus the symmetry of the conic sections implies that *the surface itself has local mirror symmetry in two perpendicular planes.* We can now derive Euler's Formula and deduce that these two perpendicular planes of symmetry are in fact the *same* planes that yield the maximum and minimum curvatures, i.e., these local mirror symmetry directions are the same as the principal directions.

Refining our coordinate system, we now align the x and y axes with these symmetry directions. Since the Eq. (4.3) is now invariant under the reflections $x \mapsto -x$ and $y \mapsto -y$ it follows that $b = 0$, and the local equation of the surface therefore becomes

$$z \asymp ax^2 + cy^2. \tag{4.4}$$

To find the geometric meaning of the coefficients a and c we now refer back to Fig. 4.3 and view it as depicting the intersection of Π_φ with S: the curve $\mathscr{C}$ is now $\mathscr{C}_\varphi$, and the tangent $\mathscr{T}$ is now the intersection of the tangent plane T_p with Π_φ, and the deviation σ of the curve from its tangent is now simply the height z of the curve above the tangent plane.

Let $\varphi = 0$ correspond to the x-axis, and let $\kappa_1 = \kappa(0)$ be the curvature of $\mathscr{C}_0 =$(the intersection of S with the xz-plane), having equation $z = ax^2$. Then the result (4.2) shows that $a = \frac{1}{2}\kappa_1$. In exactly the same way, defining $\kappa_2 = \kappa(\frac{\pi}{2})$ to be the curvature of the intersection curve with the yz-plane, we find that $c = \frac{1}{2}\kappa_2$. Thus (4.4) can be expressed more geometrically as

$$z \asymp \tfrac{1}{2}\kappa_1 x^2 + \tfrac{1}{2}\kappa_2 y^2. \tag{4.5}$$

Now consider Fig. 4.6, which depicts the curve $\mathscr{C}_\varphi$ for a general angle φ. [This diagram (and others to follow) assumes that the Gaussian curvature is positive, but the accompanying reasoning applies equally well to negatively curved surfaces.] If we move a distance ϵ within T_p in the direction φ, then we arrive at the illustrated point $x = \epsilon \cos\varphi, y = \epsilon \sin\varphi$. Thus inserting (4.5) into (4.2) yields

$$\kappa(\varphi) \asymp 2\left[\frac{z}{\epsilon^2}\right] \asymp 2\left[\frac{\frac{1}{2}\kappa_1(\epsilon \cos\varphi)^2 + \frac{1}{2}\kappa_2(\epsilon \sin\varphi)^2}{\epsilon^2}\right] = \kappa_1 \cos^2\varphi + \kappa_2 \sin^2\varphi,$$

proving Euler's Formula, and thereby establishing the extremal nature of the curvatures κ_1 and κ_2 associated with the orthogonal directions of local mirror symmetry.

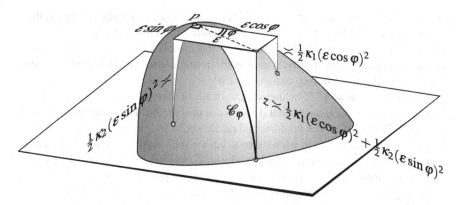

Fig. 4.6

4.3   The *Theorema Egregium* and the Gauss–Bonnet Theorem

The road to Beltrami's vindication of hyperbolic geometry was paved by Gauss. In 1816 Gauss made a discovery about the curvature of surfaces that was so profound and so unexpected that in his private notes he recorded it as "the beautiful theorem." A *decade* later,[8] in 1827, he had finally perfected his discovery to his own exacting standards, publishing the result as the centerpiece of his *Disquisitiones generales circa superficies curva* ["General Investigations of Curved Surfaces" Dombrowski 1979; Gauss 1965]. He now allowed pent-up exuberance to get the better of him, and what he had privately described as "beautiful" he now announced to the world (in Latin) as "remarkable": the *Theorema Egregium*.

The *Theorema Egregium* states, in essence, that the Gaussian curvature belongs to the *intrinsic* geometry of the surface. Intrinsic geometry means the geometry that is knowable to tiny, ant-like, intelligent (but 2-dimensional!) creatures living *within* the surface. These creatures can, for example, define a "straight line" connecting two nearby points as the shortest route within their world (the surface) connecting the two points: picture a string stretched tightly over the surface connecting the points. From there they can go on to define triangles, etc. Defined in this way, it is clear that the intrinsic geometry is unaltered when the surface is bent into quite different shapes in space, as long as distances *within* the surface are not stretched or distorted in any way. To the ant-like creatures within the surface such changes are utterly undetectable.

Under such a bending, the so-called *extrinsic* geometry (how the surface sits in space) most certainly does change, and in particular the principal curvatures κ_1 and κ_2 both change. But the *Theorema Egregium* states that the Gaussian curvature $\mathcal{K} = \kappa_1 \kappa_2$ enjoys the "remarkable" property of remaining *constant* under any

---

[8]See the fascinating chronology and insightful analysis given by Dombrowski (1979).

such bending[9] of the surface; if κ_1 doubles (for example) at a particular point, then (magically) κ_2 will be halved, thereby maintaining the same value of $\mathcal{K} = \kappa_1 \kappa_2$. Not only does this concept of $\mathcal{K}$ belong to the intrinsic geometry of the surface, but, as we shall explain shortly, it is directly *accessible* to the intelligent creatures living within the surface: they can measure it!

Although the year 1827 witnessed the death of Beethoven, the appearance of the *Theorema Egregium* meant that it also witnessed the birth of modern differential geometry. The proof of this fundamental result would take us too far afield, and we shall therefore *assume*[10] it in what follows, but we shall do so in the form of another important and beautiful result, called the (local) *Gauss–Bonnet Theorem*. Before we state this theorem for a general surface, we discuss its precursor on the sphere.

On a curved surface, the equivalent of a straight line segment connecting two points is the shortest path within the surface that connects the points: this is called a *geodesic*. As we have said, this is a concept that belongs to the *intrinsic* geometry of the surface. On the sphere the geodesics are the great circles. If on this sphere we construct a *geodesic triangle* by connecting three points with geodesics, then it is clear that one of the fundamental laws of Euclidean geometry breaks down: the interior angles add up to *more* than π.

To quantify this departure from Euclidean geometry, we introduce the *angular excess*, defined to be the amount $\mathcal{E}$ by which the angle sum exceeds π:

$$\mathcal{E} \equiv (\text{angle sum}) - \pi.$$

Since intelligent creatures within the surface are able to construct such triangles and measure the angles within them, $\mathcal{E}$ belongs to the intrinsic geometry.

As a concrete example, consider the case where two of the vertices are on the equator, and the third is at the north pole, the angle there being θ. See Fig. 4.7a. Since both angles at the equator are $(\pi/2)$, we see that $\mathcal{E} = \theta$. We also see that the area $\mathcal{A}$ of this triangle is a fraction $(\theta/2\pi)$ of the northern hemisphere, and so if the radius of the sphere is R, then $\mathcal{A} = \theta R^2$. Thus,

$$\mathcal{E} = \frac{1}{R^2}\mathcal{A}. \tag{4.6}$$

In 1603 the English mathematician Thomas Harriot discovered that this relationship holds for *any* geodesic triangle on the sphere. Harriot's beautiful proof can be found in Needham (1997, p. 278), Penrose (2005, p. 44), or Stillwell (2010, p. 350).

---

[9]The word "bending" implies continuous deformation, but this is not actually required by the theorem: there do exist isometric mappings that cannot be carried out via continuous deformation, but which nevertheless preserve the curvature by virtue of the theorem. See Aleksandrov (1969).

[10]For the classical calculational proof, see the excellent book of Banchoff (Banchoff and Lovett 2010, p. 247); for other calculational proofs, see (Berger 2003, pp. 105–111); for a compact approach using differential forms, see O'Neill (2006). In our forthcoming book, *Visual Differential Geometry*, we shall provide a simple *geometric* proof.

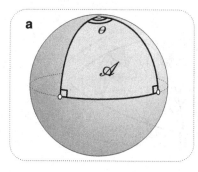
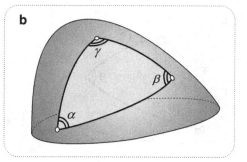

Fig. 4.7

The Gauss–Bonnet Theorem, as originally stated by Gauss in the *Disquisitiones generales*, is a stunning generalization of this result to a geodesic[11] triangle Δ on a general curved surface, illustrated in Fig. 4.7b. It says that the angular excess of such a triangle is simply the *total curvature* inside it:

$$\mathscr{E}(\Delta) = \alpha + \beta + \gamma - \pi = \iint_\Delta \mathscr{K} \, d\mathscr{A}. \tag{4.7}$$

In the case of the sphere, we see that $\kappa_1 = (1/R) = \kappa_2$, and therefore $\mathscr{K} = \kappa_1\kappa_2 = 1/R^2$. Thus (4.7) yields Harriot's formula (4.6) as a very special case.

The Gauss–Bonnet Theorem and the *Theorema Egregium* are closely connected. First we observe that the Gauss–Bonnet Theorem immediately implies the *Theorema Egregium*. For if the geodesic triangle of area $\mathscr{A}$ is shrunk down towards a point at which the Gaussian curvature is $\mathscr{K}$, then the Gauss–Bonnet Theorem implies that

$$\mathscr{E} \asymp \mathscr{K}\mathscr{A}. \tag{4.8}$$

And since both $\mathscr{E}$ and $\mathscr{A}$ are determined solely by the intrinsic geometry of the surface, it follows from this fundamental formula that $\mathscr{K} \asymp (\mathscr{E}/\mathscr{A})$ is intrinsic too: it is a quantity that can be measured by intelligent creatures living within the surface.

Conversely, if we are granted the *Theorema Egregium* in the form of (4.8), we can recover the full Gauss–Bonnet Theorem. The key fact is that the angular excess is *additive*. In Fig. 4.8a a geodesic segment [dashed] has been drawn from one vertex of Δ to an arbitrary point on the opposite edge, thereby splitting Δ into two geodesic subtriangles, Δ_1 and Δ_2. Observing that $\beta_1 + \alpha_2 = \pi$, we find that

$$\mathscr{E}(\Delta_1) + \mathscr{E}(\Delta_2) = [\alpha + \beta_1 + \gamma_1 - \pi] + [\alpha_2 + \beta + \gamma_2 - \pi] = \alpha + \beta + \gamma_1 + \gamma_2 - \pi,$$

---

[11]In 1865 Bonnet generalized the formula to non-geodesic triangles, hence the name of the theorem.

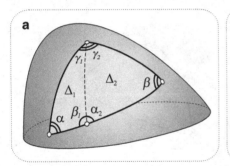
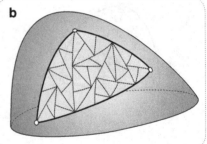

Fig. 4.8

and therefore

$$\mathscr{E}(\Delta) = \mathscr{E}(\Delta_1) + \mathscr{E}(\Delta_2).$$

These subtriangles may then be subdivided in their turn, and so on and so forth, yielding Fig. 4.8b, and the additive property ensures that $\mathscr{E}(\Delta) = \sum \mathscr{E}(\Delta_i)$. As the subdivision becomes finer and finer, the curvature varies less and less within each Δ_i, approaching the constant value $\mathscr{K}_i$, and in this limit (4.8) yields $\mathscr{E}(\Delta) = \sum \mathscr{K}_i \mathscr{A}_i$, and so we recover the Gauss–Bonnet Theorem, (4.7).

4.4 The Tractrix and the Pseudosphere

We now return to the history of the discovery of hyperbolic geometry. As early as 1766,[12] Johann Heinrich Lambert discovered a fundamental equivalent of the Hyperbolic Axiom (4.1): the angles in a triangle add up to *less* than π, and in fact *the angular excess of a triangle is a negative multiple of its area*: $\mathscr{E}(\Delta) = K\mathscr{A}(\Delta)$, where K is a negative constant. Since $\mathscr{E}$ is dimensionless, and $\mathscr{A}$ has dimensions of [length]^2, it follows that K has dimensions of $1/$[length]^2. Thus there exists a length R such that $K = -(1/R^2)$, and so Lambert's result can be written

$$\mathscr{E}(\Delta) = -\frac{1}{R^2}\mathscr{A}(\Delta). \tag{4.9}$$

Note the striking similarity to the result (4.6) on the sphere, but now with a fundamental difference: the minus sign. Lambert went so far as to say that it was as though the triangle were drawn on a sphere of *imaginary* radius iR.[13]

---

[12]Published posthumously, in 1786.

[13]This was an insight centuries ahead of its time: see Penrose (2005, §18.4).

Lambert's result was later rediscovered by Gauss (along with other results of hyperbolic geometry) but he lacked the courage or the conviction to publish his discoveries prior to Lobachevsky and Bolyai. And yet, in private, Gauss sometimes seems to have believed that hyperbolic geometry might actually exist, and he went so far as to say (see Rosenfeld 1988, p. 215) that he wished that it might apply to the real world. This was prophetic, for Einstein's 1915 discovery of General Relativity revealed that this is indeed the case, though the deviation from Euclid's geometry varies in both type and intensity from place to place, from time to time, and from direction to direction, according to the distribution of matter and energy.

Lambert's result (4.9) should have, by all rights, struck Gauss as familiar, but he failed to recognize a connection between hyperbolic geometry and his own work on differential geometry, and that happy task instead fell to our hero, Beltrami. He realized that it would follow from Gauss's result (4.7) that geodesic triangles constructed within a surface would automatically obey the defining law of hyperbolic geometry, $\mathscr{E}(\Delta) = -(1/R^2)\mathscr{A}(\Delta)$, precisely if that surface had *constant negative curvature*, $\mathscr{K} = -(1/R^2)$.

4.4.1   Construction of the Tractrix and the Pseudosphere

In fact there do exist surfaces possessing constant negative curvature $\mathscr{K}$—Beltrami called such surfaces *pseudospherical*—and all surfaces that share the same constant negative value of $\mathscr{K}$ possess the same intrinsic geometry. To begin to understand hyperbolic geometry, it is therefore sufficient to examine *any* such surface. For Beltrami's purposes, and ours, the simplest one is called the *pseudosphere*.

Try the following experiment. Take a small heavy object, such as a paperweight, and attach a length of string to it. Now place the object on a table and drag it by moving the free end of the string along the edge of the table. You will see that the object moves along a curve like that in Fig. 4.9, where the Y-axis represents the edge of the table. This curve is called the *tractrix*,[14] and the Y-axis (which the curve approaches asymptotically) is called the *axis*. The tractrix was first investigated by Newton, in 1676.

If the length of the string is R, then it follows that the tractrix has the following geometric property: *the segment of the tangent from the point of contact to the Y-axis has constant length R.* This was Newton's definition of the tractrix.

Returning to Fig. 4.9, let σ represent arc length along the tractrix, with $\sigma = 0$ corresponding to the starting position $X = R$ of the object we are dragging. Just as the object is about to pass through (X, Y), let dX denote the infinitesimal[15] change

---

[14]Same etymology as "tractor," which drags things.

[15]Here we use "infinitesimal" as a convenient shorthand for a longer description in terms of ultimate equality.

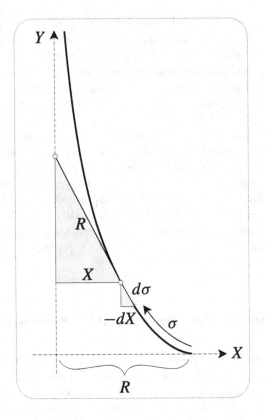

Fig. 4.9

in X that occurs while the object moves a distance $d\sigma$ along the tractrix. From the ultimate similarity of the illustrated triangles, we deduce that

$$\frac{-dX}{d\sigma} = \frac{X}{R} \quad \Longrightarrow \quad X = R\,e^{-\sigma/R}. \tag{4.10}$$

The *pseudosphere* of radius R may now be simultaneously defined and constructed as the surface obtained by rotating the tractrix about its axis. Remarkably, this surface was investigated as early as 1693 (by Christiaan Huygens), two centuries prior to its catalytic role in the acceptance of hyperbolic geometry, and the constancy of its curvature was already known to Minding in 1839.

4.4.2 The Constant Curvature of the Pseudosphere

For Beltrami's interpretation of hyperbolic geometry to work, the essential result is that the pseudosphere does indeed have constant negative Gaussian curvature. Next

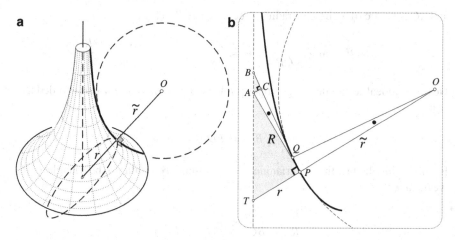

Fig. 4.10

we provide a simple geometric proof[16] of this fact. More precisely, we will use the *extrinsic* definition of $\mathcal{K}$ as the product of the principal curvatures to show that

The pseudosphere of radius R has constant curvature $\mathcal{K} = -(1/R^2)$.

At the conclusion of this essay will provide a second, *intrinsic* demonstration of this crucial fact.

Let r and $\tilde{r}$ be the two principal radii of curvature of the pseudosphere of radius R. As with any surface of revolution, it follows by symmetry that

$$\tilde{r} = \textit{radius of curvature of the generating tractrix,}$$

$$r = \textit{the segment of the normal from the surface to the axis,}$$

as illustrated in Fig. 4.10a. The problem of determining the Gaussian curvature

$$\mathcal{K} = -\frac{1}{r\,\tilde{r}}$$

is thereby reduced to a problem in plane geometry, which is solved in Fig. 4.10b.

By definition, the tractrix in this figure has tangents of constant length R. At the neighboring points P and Q, Fig. 4.10b illustrates two such tangents, PA and QB, containing angle •. The corresponding normals PO and QO therefore contain the same angle •. Note that AC has been drawn perpendicular to QB.

Now let's watch what happens as Q coalesces with P, which itself remains fixed. In this limit, O is the center of the circle of curvature, PQ is an arc of this circle,

---

[16]This proof was first published in Needham (1997).

and AC is an arc of a circle of radius R centered at P. Thus,

$$\tilde{r} \asymp OP \quad \text{and} \quad \frac{PQ}{OP} \asymp \bullet \asymp \frac{AC}{R} \quad \Longrightarrow \quad \frac{AC}{PQ} \asymp \frac{R}{\tilde{r}}.$$

Next we appeal to the defining property $PA = R = QB$ of the tractrix to deduce that

$$BC \asymp PQ.$$

Finally, using the fact that the triangle ABC is ultimately similar to the triangle TAP, we deduce that

$$\frac{r}{R} \asymp \frac{AC}{BC} \asymp \frac{AC}{PQ} \asymp \frac{R}{\tilde{r}}.$$

Behold!

$$\mathcal{K} = -\frac{1}{r\,\tilde{r}} = -\frac{1}{R^2}.$$

4.4.3   A Conformal Map of the Pseudosphere

The abstract hyperbolic geometry discovered by Lobachevsky and Bolyai is understood to take place in a *hyperbolic plane* that is exactly like the Euclidean plane, *except* that lines within it obey the Hyperbolic Axiom (4.1).

The constant negative curvature of the pseudosphere ensures that it faithfully embodies local consequences of this axiom, but the pseudosphere will not do as a model of the *entire* hyperbolic plane, because it departs from the Euclidean plane in two unacceptable ways. First, the pseudosphere is akin to a cylinder instead of a plane: a loop in the plane can be shrunk down to a point, but a loop that wraps around the axis of the pseudosphere is like a circular cross section of a cylinder, which *cannot* be shrunk to a point. Second, a geodesic segment on the pseudosphere cannot be extended indefinitely in both directions: we hit the rim.[17]

Beltrami recognized these obstacles, and he overcame them both, in one fell swoop, by constructing a conformal map of the pseudosphere. As a first step towards this map, Fig. 4.11a introduces a natural coordinate system (x, σ) on the pseudosphere.

---

[17]To make matters worse, it turns out [as Hilbert later discovered in 1901] that such a rim is an *essential* feature of *all* surfaces of constant negative curvature—not intrinsically, but by virtue of trying to force the surfaces to fit inside ordinary Euclidean three-space.

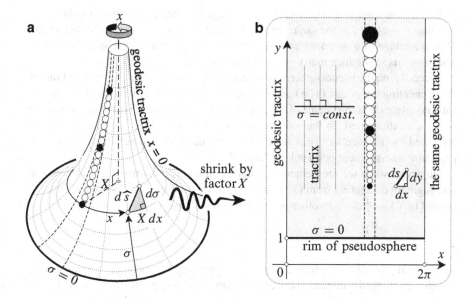

Fig. 4.11

The first coordinate x measures angle around the axis of the pseudosphere, say restricted to $0 \le x < 2\pi$. The second coordinate σ measures arc length along the tractrix curves (called *generators*) that make up the pseudosphere (as in Fig. 4.9a). Thus the curves $x =$ const. are the tractrix generators of the pseudosphere [note that these are clearly geodesics], and the curves $\sigma =$ const. are circular cross sections of the pseudosphere [note that these are clearly *not* geodesics]. Since the radius of such a circle is the same thing as the X-coordinate in Fig. 4.9a, it follows from (4.10) that

The radius X of the circle $\sigma =$ const. passing through the point (x, σ) is given by
$X = R\,e^{-\sigma/R}$.

In our map, let us choose the angle x as our horizontal axis, so that the tractrix generators of the pseudosphere are represented by vertical lines. See Fig. 4.11b. Thus a point on the pseudosphere with coordinates (x, σ) will be represented in the map by a point with Cartesian coordinates (x, y), which we will soon think of as the complex number $z = x + iy$.

If our map were not required to be special in any way, then we could simply choose $y = y(x, \sigma)$ to be an arbitrary function of x and σ. But suppose instead that our map is required to *preserve angles*: such a map is called *conformal*. Thus an infinitesimal triangle on the pseudosphere is mapped to a *similar* infinitesimal triangle in the map, and more generally it follows that any small shape on the pseudosphere looks the same (only bigger or smaller) in the map. Having decided upon such a conformal map, we will now discover that there is (virtually) no freedom in the choice of the y-coordinate.

Firstly, the tractrix generators x = const. are orthogonal to the circular cross sections σ = const., so the same must be true of their images in our conformal map. Thus the image of σ = const. must be represented by a horizontal line y = const., and from this we deduce that $y = y(\sigma)$ must be a function solely of σ.

Secondly, on the pseudosphere consider the arc of the circle σ = const. (of radius X) connecting the points (x, σ) and $(x + dx, \sigma)$. By the definition of x, these points subtend angle dx at the center of the circle, so their separation on the pseudosphere is $X\, dx$, as illustrated. In the map, these two points have the same height and are separated by distance dx. Thus in passing from the pseudosphere to the map, this particular line-segment is shrunk by factor X.

However, since we are demanding that our map be conformal, an infinitesimal line-segment emanating from (x, σ) in *any* direction must be multiplied by the *same* factor $(1/X) = \frac{1}{R}e^{\sigma/R}$. In other words, the so-called *metric* is

$$d\hat{s} = X\, ds,$$

where $d\hat{s}$ represents the distance between neighboring points on the pseudosphere, and ds represents the distance between the corresponding points in the map.

Thirdly, consider the uppermost black disc on the pseudosphere shown in Fig. 4.11a. Think of this disc as infinitesimal, say of diameter ϵ. In the map, it will be represented by *another disc*, whose diameter (ϵ/X) may be interpreted more vividly as the angular width of the original disc as seen by an observer at the same height on the pseudosphere's axis. Now suppose we repeatedly translate the original disc towards the pseudosphere's rim, moving it a distance ϵ each time. Figure 4.11a illustrates the resulting chain of touching, congruent discs. As the disc moves down the pseudosphere, it recedes from the axis, and its angular width as seen from the axis therefore diminishes. Thus the image disc in the map appears to gradually shrink as it moves downward, and the equal distances 8ϵ between the successive black discs certainly do not appear equal in the map.

Having developed a feel for how the map works, let's actually calculate the y-coordinate corresponding to the point (x, σ) on the pseudosphere. From the above observations (or directly from the requirement that the illustrated triangles be similar) we deduce that

$$\frac{dy}{d\sigma} = \frac{1}{X} = \frac{1}{R}e^{\sigma/R} \quad \Longrightarrow \quad y = e^{\sigma/R} + \text{const.}$$

The standard choice of this constant is 0, so that

$$y = e^{\sigma/R} = (R/X). \tag{4.11}$$

Thus the entire pseudosphere is represented in the map by the shaded region lying above the line $y = 1$ (which itself represents the pseudosphere's rim), and the metric associated with the map is

$$d\hat{s} = \frac{R\,ds}{y} = \frac{R\,\sqrt{dx^2 + dy^2}}{y}. \tag{4.12}$$

For future use, also note that an infinitesimal rectangle in the map with sides dx and dy represents a similar infinitesimal rectangle on the pseudosphere with sides $(R\,dx/y)$ and $(R\,dy/y)$. Thus the apparent area $dx\,dy$ in the map is related to the true area $d\mathscr{A}$ on the pseudosphere by

$$d\mathscr{A} = \frac{R^2\,dx\,dy}{y^2}. \tag{4.13}$$

4.5 Beltrami's Hyperbolic Plane: Geodesics and Isometries

We now have a conformal map of the *(1) cylinder-like, (2) rimmed,* pseudosphere: $\{(x, y) : 0 \leqslant x < 2\pi, y \geqslant 1\}$. To instead create a map of an infinite hyperbolic plane, Beltrami knew that he must remove both of these adjectives. Note that while different choices of R yield quantitatively different geometries, they are all qualitatively the same, and *in this section only we shall make the conventional choice* $R = 1$.

To remove the "cylinder-like" adjective, imagine painting a wall with a standard cylindrical paint roller (of unit radius). After one revolution you have painted a strip of wall of width 2π, and every point on the surface of the roller has been mapped to a unique point within this strip of flat wall. To paint the entire wall, you can simply keep on rolling! Now imagine that our paint roller instead takes the form of a pseudosphere. To make it fit onto the flat wall you must first stretch out its surface, according to the metric (4.12), but then, just as before, you can keep on rolling. If a particle moves along a horizontal line on the wall, for example, the corresponding particle on the pseudosphere goes round and round the horizontal circle $\sigma = $ const. The "cylinder-like" adjective has been successfully removed[18] and we now have map $\{(x, y) : -\infty < x < \infty, y \geqslant 1\}$.

The conformal map solves our second problem of the pseudosphere's rim, with equal ease. On the left of Fig. 4.12 is the image of a particle moving down the pseudosphere along a tractrix. Of course on the pseudosphere the journey is rudely interrupted at some point $\hat{p}$ on the rim ($\sigma = 0$), corresponding to a point p on the line $y = 1$. But in the map this point p is just like any other, and there is absolutely nothing preventing us from continuing all the way down to the point q on $y = 0$, with the true distances $d\hat{s}$ continuing to be given by $d\hat{s} = \frac{ds}{y}$.

---

[18]Stillwell notes (Stillwell 1996) that this was perhaps the first appearance in mathematics of what topologists now call a *universal cover*.

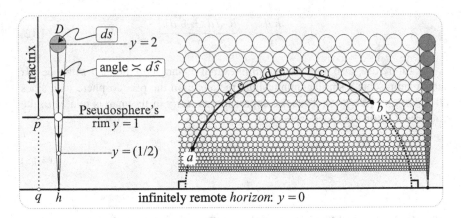

Fig. 4.12

Why stop at q? The answer is that the particle will never even get that far, because q is infinitely far from p! Consider the small disc D of diameter ds on the line $y = 2$ shown on the left of Fig. 4.12. Its true size on the pseudosphere is $d\hat{s} = \frac{ds}{y}$, and this is ultimately equal to the illustrated angle it subtends at the point h directly below it on the line $y = 0$. Now imagine D moving down the pseudosphere at steady speed. Its apparent size in the map must shrink so that its subtends a constant angle at h. In the map it hits $y = 1, \ldots$ and keeps on going!

Assuming it took one unit of time to go from $y = 2$ to $y = 1$, then in the next unit of time it will reach $y = (1/2)$, then $y = (1/4)$, etc. Thus, viewed within the map, each successive unit of time only halves the distance from $y = 0$, and therefore D will never reach it. [An appropriate name for this phenomenon might be "Zeno's Revenge"!]

At last, we now possess a concrete model of the *hyperbolic plane*: the entire shaded half-plane $y > 0$, with metric $d\hat{s} = \frac{ds}{y}$. The points on the real axis are infinitely far from ordinary points and are not (strictly speaking) considered part of the hyperbolic plane. They are called *ideal points*, or *points at infinity*. The complete line $y = 0$ of points at infinity is called the *horizon*.

Although Beltrami discovered this map in 1868 (anticipating Poincaré by 14 years), it is now universally known as the *Poincaré half-plane*. But since one of our main aims is to stress Beltrami's contributions, in the remainder of this essay we shall doggedly refer to this map as the *Beltrami–Poincaré half-plane*.

Let us attempt to make the metric of this model more vivid. On the far right of Fig. 4.12 is a vertical string of touching circles of equal hyperbolic size ϵ, as in Fig. 4.11. To the left of this we have filled part of the hyperbolic plane with such circles, *all of equal hyperbolic diameter ϵ*. Thus the hyperbolic length of any curve can be gauged by the number of circles (and fractions of circles) it intercepts, multiplied by ϵ. This makes it clear that the shortest route from a to b is the one that intercepts the smallest number of circles, and which therefore has the approximate shape shown.

Though we shall not prove it here,[19] the figure illustrates the first of several miracles: the exact form of such a geodesic is a perfect *semicircle that meets the horizon at right angles*. [The only geodesics that are not of this form are the vertical lines (extending the tractrix generators), but even these may be viewed as a limiting case in which the radius of the semicircle tends to infinity.] Note that the truth of the Hyperbolic Axiom (4.1) is now self-evident; indeed, there are *infinitely* many such semicircles through a given point p that do not meet a particular given one L.

There are more miracles associated with the Beltrami–Poincaré half-plane map. Since it is conformal (by construction) infinitesimal circles on the pseudosphere are represented by infinitesimal circles in the map. But in fact a circle of *any* size maps to a perfect circle, though the center does *not* map to the center of the corresponding circle in the map.

While Beltrami was the first to discover the conformal maps, Poincaré made other wonderful discoveries that were without precedent. Our third and final miracle is an example of this. In 1882 Poincaré found[20] that the hyperbolic *isometries* [rigid motions that preserve distances] have a remarkably simple form, but only when viewed in the *complex plane*. Thus the map is now thought of as the upper half of the complex plane, and the point (x, y) is now taken to represent the complex number $z = x + iy$.

A rotation of the pseudosphere through angle θ about its axis is represented in the complex plane by $z \mapsto z + \theta$. Also, each successive equal downward translation of D in Fig. 4.12 is represented by a contraction by factor 2 centered at h; more general translations are likewise represented by dilations $z \mapsto rz$ (r real). But such isometries are actually insufficient to generate the most general motions within the hyperbolic plane: we are missing rotations.

A rotation of ϕ about a point p can be built out of two successive reflections in lines that intersect at p at angle $(\phi/2)$. Poincaré discovered that reflection in a hyperbolic "straight line" is in fact *geometric inversion* (see Needham 1997, Chap. 3) in the corresponding semi-circular representation of the geodesic (such as in Fig. 4.12). He was thereby able to deduce the lovely and important fact that the most general isometry of the Beltrami–Poincaré half-plane is given by a *Möbius transformation* (with real coefficients):

$$z \longmapsto \frac{az + b}{cz + d}, \qquad (ad - bc) > 0.$$

---

[19]See (Needham 1997, Chap. 6) or (Stillwell 1992, Chap. 4) for the complete story.

[20]See Stillwell (1996), Needham (1997, Chap. 6), Stillwell (1992, Chap. 4), and Stillwell (2010).

That this is closely connected to the previous miracle can be seen in the fact that Möbius transformations are not only conformal, but are known to preserve circles of all sizes.[21]

4.6 Parallel Transport

4.6.1 Parallel Transport in a General Surface

Imagine yourself standing in the midst of a vast desert that is utterly featureless, except for two objects: at your feet an archer's arrow lies on the ground, and on the horizon an obelisk rises from the hot sand. You are now given a task, but no equipment to carry it out. Your task is to transport the arrow to the obelisk while always keeping it parallel to the original direction in which you found it on the ground.

The solution is deceptively simple. You pick up the arrow, being careful not to disturb its direction, set your sights on the obelisk, and begin walking towards it in a straight line. As you walk in this straight line, always keeping eyes front on the obelisk, you ensure that the arrow maintains a *constant angle* with the direction in which you walk. When you arrive at the obelisk, you set the arrow down, confident that it is parallel to its original direction. You have just *parallel-transported* the arrow to the obelisk.

But your confidence is misplaced! The next day you are instructed to repeat this feat, starting at the same place and with a second arrow that points in the same direction as the first. This time you realize that you had overlooked an oasis off to the right of the obelisk, and you decide to quench your thirst along the route. You parallel-transport the arrow to the oasis, set it down on the ground, drink and rest. Refreshed, you pick up the arrow and complete your journey to the obelisk. But when you set the second arrow down next to the first, you are shocked to discover that it points in a slightly different direction: *parallel transport depends on the route taken!*

Let $\mathscr{R}$ denote the angle of rotation from the first vector [direct route] to the second [via oasis]. This rotation $\mathscr{R}$ cannot depend upon the direction of the vector you transported, for if you had simultaneously carried two vectors along either route, the angle between them would have remained constant by virtue of each of them maintaining constant angle with your direction of motion. In fact, this direction-independent rotation $\mathscr{R}$ measures the total Gaussian curvature contained in the part of the Earth's surface that is enclosed by your two routes.

To begin to understand this, consider Fig. 4.13, which depicts an extreme form of the above experiment. Since $\mathscr{R}$ is direction-independent, we are free to choose any

---

[21]See (Needham 1997, Chap. 3).

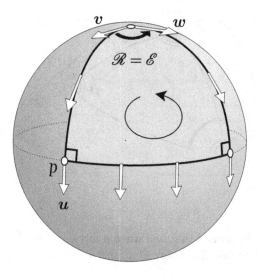

Fig. 4.13

initial vector we like: we choose **u** at p to point due south. If we parallel-transport **u** due north from p, maintaining zero angle with the "straight line" (geodesic) along which we carry it, then when we arrive at the north pole we have **v**. But if we instead transport **u** along the equator and then head north to the pole, we obtain the quite different vector **w**. To put this differently, if we start with **v** at the north pole, and parallel-transport it counterclockwise round the geodesic triangle, say Δ, then it returns to the north pole as **w**. Note that it has undergone a rotation $\mathscr{R}$ in the *same direction* as the direction of transport.

If we had transported the vector on a surface of *negative* curvature, then the rotation would have been *opposite* to the direction of transport. We strongly encourage you to verify this empirically by stretching pieces of string over a physical surface and then parallel-transporting a vector along these geodesics. You may easily equip your laboratory at the grocery store, where suitable fruits and vegetables with patches of negative curvature are readily available.

Comparison of Fig. 4.13 with our earlier discussion of Fig. 4.7a reveals that the angle of rotation is none other than the angular excess, which is indeed direction-independent, and which by virtue of (4.7) equals the total Gaussian curvature within:

$$\mathscr{R}(\Delta) = \mathscr{E}(\Delta) = \iint_{\Delta} \mathscr{K} \, d\mathscr{A}.$$

That this is a completely general result is proved in Fig. 4.14, which depicts a general triangle on a general surface. Here we have chosen **v** to point along the first side, so it remains tangent as it is transported to the next vertex, where it therefore makes angle $* = \pi - \beta$ with the second side. It maintains this angle $*$ as it is

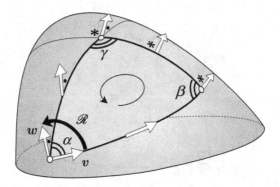

Fig. 4.14

transported along the second side, and when it arrives at the next vertex it makes angle • with the third side. Since we can see that • + * = γ, we deduce that

$$\bullet = \gamma - * = \beta + \gamma - \pi.$$

Finally, as the vector is transported along the third side it maintains angle • with that side, returning home as w, having undergone a net rotation of

$$\mathscr{R}(\Delta) = \alpha + \bullet = \alpha + \beta + \gamma - \pi = \mathscr{E}(\Delta).$$

The equivalence of these two measures of curvature proves that $\mathscr{R}$ must be *additive*, so that in Fig. 4.8a we have

$$\mathscr{R}(\Delta) = \mathscr{R}(\Delta_1) + \mathscr{R}(\Delta_2).$$

But it is both more satisfying and more enlightening to see this *directly* (without reference to $\mathscr{E}$) by transporting a vector round Δ_1 and then round Δ_2. We strongly encourage you to carry out this instructive exercise. [Hint: what happens along the dashed boundary between Δ_1 and Δ_2?] A simple extension of this reasoning shows that even if Δ is a geodesic polygon with many sides, rather than a mere triangle, $\mathscr{R}(\Delta)$ continues to measure the total curvature inside.

Lastly we observe that if we wish to parallel-transport a vector along a loop L that is *not* geodesic, then we may do so by approximating L with a geodesic polygon with many short geodesic sides, then taking the limit as the length of each side tends to zero. [We will provide a concrete example of this construction in the next section.] Thus in this case, too, $\mathscr{R}(L)$ measures the total curvature inside L.

This provides a new, more powerful, way of thinking about the curvature $\mathscr{K}(p)$ at a point p. If L is a small loop surrounding p, then

$$\mathscr{K}(p) = \textit{rotation per unit area, as L shrinks to p.}$$

This opens the door to Riemann's brilliant generalization of curvature from 2-dimensional surfaces to n-dimensional spaces.

We cannot go into detail here, but let us sketch the idea, deliberately sweeping some complications under the rug.[22] Within this n-dimensional space, let L be a very small parallelogram of area $\mathscr{A}$ with edge vectors l_1 and l_2. Let v be a unit vector in an arbitrary direction; remember, we are no longer trapped in two dimensions, so v is now free to point *out* of the plane of L. After parallel-transporting v round L, we return it as w. The net effect is that v has undergone a rotation in the direction $R = w - v$ [the vector from the tip of v to the tip of w] through angle $\mathscr{R} = |R|$ (because an infinitesimal rotation will move the tip of a unit vector a distance equal to the angle of that rotation).

Since R depends on all three vectors, we may write it as a function $R(l_1, l_2, v)$. Remarkably, it can be shown that R is a *linear* function of all three vectors:

$$R(l_1, l_2, au + bv) = a R(l_1, l_2, u) + b R(l_1, l_2, v),$$

and likewise for the first two slots. Therefore R is what mathematicians call a *tensor*; it is in fact the famous *Riemann tensor.*

While $(\mathscr{R}/\mathscr{A})$ (rotation per unit area) continues to measure curvature, there are now *many* such curvatures, depending on the orientation of L and v. In fact it turns out that R is now characterized by an array of $\frac{1}{12}n^2(n^2-1)$ numbers, called curvature *components*. Thus on a surface ($n = 2$) there is only a single component $\mathscr{K}$, while in Einstein's curved spacetime ($n = 4$) the gravitational field has 20 components.

4.6.2   Parallel Transport in the Hyperbolic Plane

We end by applying the concept of parallel transport to the Beltrami–Poincaré half-plane, thereby providing a simple *intrinsic* geometric demonstration of its constant negative curvature.

On the pseudosphere of radius R, consider the rectangle $abcd$ (traced counterclockwise) bounded by the vertical segments ad and bc of geodesic tractrix generators (θ being the angle from the first to the second) together with the non-geodesic horizontal circular arcs ab, cd. See the left side of Fig. 4.15. As illustrated, let us parallel-transport a vector round $abcd$ to discover the total curvature within.

The right side of Fig. 4.15 depicts the conformal image in the Beltrami–Poincaré model: $abcd$ is mapped to the rectangle with vertices $A = (x, y)$, $B = (x + \theta, y)$, $C = (x + \theta, Y)$, $D = (x, Y)$. Thus, using (4.13), the area $\mathscr{A}$ of the rectangle $abcd$ on the pseudosphere is

---

[22]For a pukka treatment, see (Penrose 2005, Chap. 14).

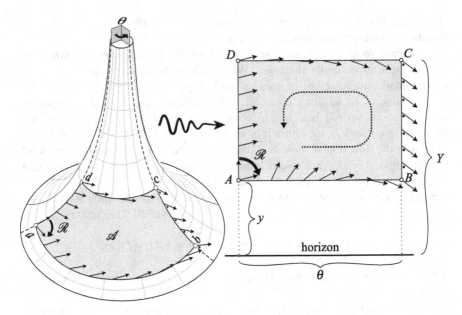

Fig. 4.15

$$\mathscr{A} = \int_{x=0}^{x=\theta} \int_{y}^{Y} \frac{R^2 dx\, dy}{y^2} = R^2\theta \left[\frac{1}{y} - \frac{1}{Y} \right]. \qquad (4.14)$$

At a we have chosen an initial vector pointing up the pseudosphere, along ad. As we parallel-transport it along ab, it rotates clockwise relative to the direction of motion; along bc it maintains constant angle • with the direction of motion (because it is geodesic); along cd it rotates counterclockwise relative to the direction of motion, *but not as much as it did on* ab; finally, it maintains constant angle with the geodesic da, returning to a having undergone a negative net rotation of $\mathscr{R}$.

Because the Beltrami–Poincaré map is conformal, when the vector is transported around $ABCD$ it undergoes the *same* net rotation $\mathscr{R}$. But, as we now explain, the crucial advantage of the map is that it enables us to *see* what this rotation actually is.

Divide the non-geodesic horizontal segment AB of (Euclidean) length θ into n small segments of length (θ/n). Next, approximate these segments with geodesic segments (see Fig. 4.16): recall that these are arcs of circles centered on the horizon. Let ϵ be the angle that each such arc subtends on the horizon, as illustrated.

As the initially vertical *Start* vector is parallel-transported along the first geodesic segment, its angle with that segment remains constant, and it therefore rotates through angle $-\epsilon$. Likewise for each successive segment, so that after all n segments have been traversed the total rotation from *Start* to *Finish* is $-n\epsilon$. But since

$$r\epsilon \asymp \frac{\theta}{n} \qquad \text{and} \qquad r \asymp y,$$

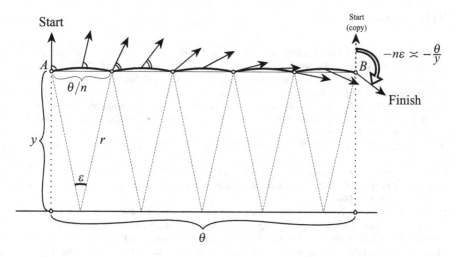

Fig. 4.16

we deduce that the total angle through which the vector is rotated in the map is

$$\mathscr{R}_{AB} \asymp -n\epsilon \asymp -\frac{\theta}{r} \asymp -\frac{\theta}{y}.$$

The same reasoning yields $\mathscr{R}_{CD} = (\theta/Y)$. And since the vector does not rotate along either BC or DA, we deduce that the net rotation upon returning to A is

$$\mathscr{R} = \mathscr{R}_{AB} + \mathscr{R}_{CD} = -\frac{\theta}{y} + \frac{\theta}{Y} = \left[-\frac{1}{R^2}\right]\mathscr{A},$$

by virtue of (4.14). Thus,

$$Rotation\ per\ unit\ area = -\frac{1}{R^2}.$$

The fact that this answer is independent of the size, shape, and location of the rectangle proves that Beltrami's conformal model (unwinding and extending the pseudosphere into an infinite plane) does indeed have *constant negative intrinsic curvature* $-1/R^2$, and thus departs from Euclid's geometry in precisely the way envisioned by Lobachevsky and Bolyai in their remarkable new geometry.

Acknowledgments I am extremely grateful to Douglas R. Hofstadter and to Paul Zeitz for reading the first draft of this essay and for providing me with many insightful comments and corrections.

References

Aleksandrov, A. D. (1969). Chapter VII: Curves and surfaces. In A. D. Aleksandrov, A. N. Kolmogorov, & M. A. Lavrent'ev (Eds.), *Mathematics: Its content, methods, and meaning* (Vol. II, pp. 57–117). Translated from the Russian by S. H. Gould. Second, paperback edition. Cambridge: The M.I.T. Press.

Arnol'd, V. I. (1990). *Huygens and Barrow, Newton and Hooke. Pioneers in mathematical analysis and catastrophe theory from evolvents to quasicrystals.* Basel: Birkhäuser. Translated from the Russian by Eric J. F. Primrose.

Banchoff, T., & Lovett, S. (2010). *Differential geometry of curves and surfaces.* Natick: A.K. Peters.

Berger, M. (2003). *A panoramic view of Riemannian geometry.* Berlin: Springer.

Bloye, N., & Huggett, S. (2011). Newton, the geometer. *Newsletter of the European Mathematical Society,* (Vol. 82), 19–27

Brackenridge, J. B., & Nauenberg, M. (2002). Curvature in newton's dynamics. In I. B. Cohen, & G. E. Smith (Eds.), *The Cambridge companion to Newton.* Cambridge companions to philosophy, Chap. 3 (pp. 85–137). Cambridge: Cambridge University Press.

Chandrasekhar, S. (1995). *Newton's Principia for the common reader.* Oxford: Clarendon Press.

de Gandt, F. (1995). *Force and geometry in Newton's principia.* Princeton: Princeton University Press. Translated from the French original and with an introduction by Curtis Wilson.

Dombrowski, P. (1979). *150 years after Gauss' "Disquisitiones generales circa superficies curvas".* Astérisque (Vol. 62). Paris: Société Mathématique de France. With the original text of Gauss.

Gauss, C. F. (1965). *General investigations of curved surfaces,* (1827). Translated from the Latin and German by Adam Hiltebeitel and James Moreh ead. Hewlett: Raven Press, 1965 edition.

Guicciardini, N. (1999). *Reading the Principia. The debate on Newton's mathematical methods for natural philosophy from 1687 to 1736.* Cambridge: Cambridge University Press.

Guicciardini, N. (2009). *Isaac Newton on mathematical certainty and method.* Transformations. Cambridge: MIT Press.

Milnor, J. (1982). Hyperbolic geometry: the first 150 years. *Bulletin of the American Mathematical Society (N.S.) 6*(1), 9–24.

Needham, T. (1993). Newton and the transmutation of force. *American Mathematical Monthly, 100*(2), 119–137.

Needham, T. (1997). *Visual complex analysis.* Oxford: Clarendon Press.

Newton, I. (1967). Normals, curvature and the resolution of the general problem of tangents. In *The mathematical papers of Isaac Newton.* (Vol. I, 1664–1666, pp. 245–297). London: Cambridge University Press.

Newton, I. (1999). *The Principia: Mathematical principles of natural philosophy.* Berkeley: University of California Press, 1999 edition (first published 1687). A new translation by I. Bernard Cohen and Anne Whitman, assisted by Julia Budenz, Preceded by "A guide to Newton's *Principia*" by Cohen.

O'Neill, B. (2006). *Elementary differential geometry* (rev. 2nd ed.). Amsterdam: Elsevier Academic.

Penrose, R. (2005). *The road to reality. A complete guide to the laws of the universe.* New York: Alfred A. Knopf Inc.

Rosenfeld, B. A. (1988). *A history of Non-Euclidean geometry: Evolution of the concept of a geometric space* (Vol. 12). New York: Springer.

Stewart, G. W. (1993). On the early history of the singular value decomposition. *SIAM Review, 35*(4), 551–566.

Stillwell, J. (1992). *Geometry of surfaces.* Universitext. New York: Springer.

Stillwell, J. (1996). *Sources of hyperbolic geometry*. History of mathematics (Vol. 10) Providence: American Mathematical Society.

Stillwell, J. (2010). *Mathematics and its history* (3rd ed.). New York: Springer.

Westfall, R. S. (1980). *Never at rest. A biography of Isaac Newton*. Cambridge: Cambridge University Press.

Chapter 5
All Done by Mirrors: Symmetries, Quaternions, Spinors, and Clifford Algebras

Simon Altmann

Some form of applied aesthetics was already implicit in some of the oldest works done by humans, and mirror symmetry was its earliest manifestation. How it came about that mirror planes were first used was already suggested by George Herbert, a seventeenth century Welsh-born English poet, who celebrated the symmetry of humans in a poem called *Man*:

> Man is all symmetrie,
> Full of proportions, one limbe to another,
> And all to all the world besides:
> Each part may call the farthest, brother

It is not unreasonable to assume, indeed, that humans had a natural introduction to symmetry through the observation of their own bodies. Of course, no one is perfectly symmetric and certainly not so if internal organs are considered, but we all have a symmetry plane (also called a *mirror plane*) with respect to which we find that eyes and ears and the proximal end of limbs are equidistant. Even the earliest examples of art objects, such as the Venus of Willendorf (ca 24,000–22,000 BCE, thus much older than Lascaux or Altamira) respect such constraints (see Fig. 5.1). Going from mirrors to rotations took many centuries, but the decoration of pots with a circular cross-section made it quite natural the use of patterns repeated through the rotation by a fixed angle around a central axis or point, for which Greek pots offer numerous early examples. Later, the desire for mural and floor decoration, especially by means of mosaics or tiles, led to more complex rotational patterns, as we shall see.

Much later, in the nineteenth century, mathematicians formalized the study of rotations, in particular through the discovery of quaternions by Sir William Rowan Hamilton in 1843, which influenced profoundly not only our knowledge of physics, especially quantum mechanics (with the introduction of spinors), but

S. Altmann (✉)
Brasenose College, Oxford, UK
e-mail: simon.altmann@bnc.ox.ac.uk

R. Lupacchini and A. Angelini (eds.), *The Art of Science*,
DOI 10.1007/978-3-319-02111-9_5,
© Springer International Publishing Switzerland 2014

Fig. 5.1 *Venus of Willendorf,*
c. 24,000–22,000 BCE.
Vienna, Naturhistorisches
Museum

also our technological scope in the design of sophisticated systems like satellites
and robots. It did not take long, however, for mathematicians to become aware of
the intrinsic limitations of rotational symmetry so that mathematics went back to the
earliest used symmetry operations, mirror reflections, through the powerful algebra
invented towards the end of the nineteenth century by William Kingdon Clifford.
So, early humans were right: their simplest forms of art led mathematics eventually
to algebras where all is done by mirrors.

5.1 Mirror Symmetry

Ever since Narcissus looked at his face reflected on a pond's surface humanity has
been both fascinated and concerned about mirrors. Not without reason their use
was deprecated if not condemned in the Middle Ages as sinful: "*le miroir est le
vraie cul du diable*" was a saw that more than one mother must have repeated to a
wanton daughter. But *reflection* or *mirror planes* had nevertheless a very respectable
pedigree, especially in architecture. They are characterized by dividing an object
in two halves on the right and left of the reflection plane in such a way that
each half is the specular (or mirror) reflection of the other, whence the name of
mirror symmetry given to this situation. Though the vanity use of mirrors was not
encouraged, mirror planes in buildings have always been used to emphasize their
dignity and importance. We do not have to go back to the Parthenon as an example:
perhaps the most notorious building in the world as a centre of power, the White
House (Fig. 5.2) is just as symmetric as you might wish, except for a presumably

Fig. 5.2 James Hoban: *The White House*, 1792–1800. Washington DC

Fig. 5.3 Philip Johnson: *The Sony Building*, 1984. NY, Manhattan

later addition to the roof line on the right-hand side of the pediment. But it is not only politicians that have to exude power: AT&T, the original owners of the now Sony building in Manhattan (Fig. 5.3), required their architect to emphasize the power and strength of their company (soon however to be taken over by the Southwestern Bell Corporation). The architect was none less than Philip Johnson, who had received from Mies van der Rohe the mantle of the leading exponent of the severe undecorated *modern style* in architecture. Amazingly, Johnson's response, at 78, was to renege from his past and crown the building with a broken pediment, no doubt thus emphasizing its mirror symmetry but perhaps also providing a subconscious metaphor for his breaking decades of the stark style that he had until then embraced: this was perhaps the birth, in 1984, of *post-modern architecture*.

One of the features of mirrors that has intrigued people over the centuries is the fact that, on reflection, objects invert left and right: when we look at ourselves in a mirror the reflection of our right arm is a left one. The mirror, however, is just as symmetrical left and right as top and bottom. Why then on reflection we do not have a top and bottom inversion as well? Even the great American physicist Richard Feynman got involved in amazing contortions in order to explain this fact (Gregory 1997). Before I discuss this problem in more detail it is important to dispel a misconception about the use of one of the most important principles in physics, the *Symmetry Principle*. Although some forms of the principle were used in a fairly intuitive way for a long time, it was Pierre Curie in 1894 who produced its first

Fig. 5.4 Specular reflection

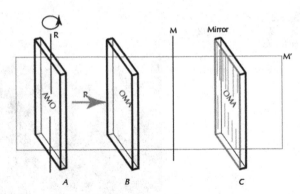

R is a rotation by 180° degrees (binary rotation).
M is a reflection plane that may be taken to coincide with the physical mirror.
M' is a reflection plane perpendicular to M.

clear account. He understood that an effect might have more symmetry than its cause, because physical systems tend to average out perturbations: a copper sphere, for instance, when heated by a gas jet along a diameter of the sphere would, if symmetry were strictly conserved, create a temperature distribution centred around that diameter but otherwise not uniform. In practice, in a very short time, the temperature of the sphere is uniform throughout it, that is, it acquires spherical symmetry, much higher than that of the jet that has caused it. Thus it is not right to expect symmetry to be simply conserved, because it can be incremented from the cause to the effect. Curie observed instead that it is the asymmetries that must be conserved: *The asymmetry of the effect must be pre-existent in its cause.* This is no more than Leibniz's principle of sufficient reason applied to asymmetries.[1]

The question that has worried many people is that the asymmetry observed in specular reflection (which exchanges right and left) does not agree with the right–left and top–bottom symmetry of the mirror, thus appearing to break the symmetry principle. But such a conclusion results from an inadequate consideration of the phenomenon, as illustrated in Fig. 5.4. In *A* we show the word AMO written on the recto of a thin Perspex box, which permits us more clearly than a sheet of paper to discriminate in the picture between recto and verso. In *B* we have rotated the box by π so that the recto now faces the mirror. Although from the mirror itself we would still see the word AMO, on the back of the box, as shown in the picture, we see OMA, which is the *mirror writing* we see reflected on the mirror (C). This is the right–left inversion, that has puzzled people for centuries but which the sagacious Hercule Poirot exploited more than once when solving his mysteries. In order to explain the effect shown in Fig. 5.4, we prove in Fig. 5.5 the following result.

- The rotation *R* followed by a reflection *M* on a mirror equals a reflection *M'* on a plane perpendicular to *M*.

---

[1]For a fuller discussion of the symmetry principle, see Altmann (1992, pp. 25–29) and Altmann (2002, p. 171).

Fig. 5.5 The symmetry elements R, M, and M' are the same as in Fig. 5.4

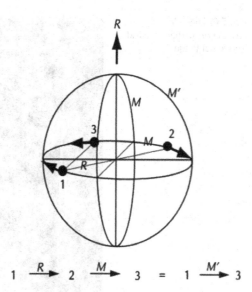

$$1 \xrightarrow{R} 2 \xrightarrow{M} 3 \;=\; 1 \xrightarrow{M'} 3$$

In order to prove this result we must first stress that the essence of symmetry operations is that they transform points on the surface of a sphere into other points on the same surface, as we do in Fig. 5.5, which confirms the statement above. What we are saying is that if we write the word AMO on a piece of paper the result of presenting it to a mirror M is the same as reflecting the word on a mirror M' perpendicular to M. This is what is called *mirror writing*, as used by Leonardo da Vinci in his notebooks in order to protect his ideas from prying eyes. It is worth mentioning two things. First that there is nothing in this result that would warrant a similar effect in the up and down direction, which worried people in the past. The second is that the apparent breach of the symmetry principle results from considering in Fig. 5.4 the mirror alone and not the binary axis R, which is in fact responsible for the right–left (specular) symmetry, which is very significant in art.

5.1.1   Interlude

5.1.1.1   Mirror Appearance

Mirrors are important in works of art as much by their presence as by their absence, the most important cases being etchings, self-portraits, and tapestries, where the mirrors are not seen, but they are also important in paintings where they are. Of the latter perhaps the two most famous are the *Arnolfini Portrait* (1434) by Jan van Eyck at the National Gallery of London and *Las Meninas* (1656) by Velázquez at the Prado Museum in Madrid. Such significance, however, as mirrors might have is probably quite different in these two pictures. In the Arnolfini the mirror, which is

Fig. 5.6 Diego Velázquez:
Las Meninas, 1656. Madrid,
Museo del Prado

a round convex one, is on the back wall at the vanishing point and reflects the back of the portrayed couple, as well as two characters not seen in the main picture.[2] It is surrounded by ten small roundels on the frame depicting scenes of the life of Jesus, and it might symbolize some sort of a heavenly world, well beyond the sub-lunar one of the room depicted in the picture. But this is mere speculation, although mirrors in fact had always some mystical connotations, especially exploited in the Kabbalah, and not unusual amongst alchemists. *Las Meninas* (Fig. 5.6), however, entails a different problem. The question here is that this magnificent picture contains a self-portrait of Velázquez, for which a mirror would normally be required by the painter although it would not appear in the picture. At one time it was thought that this had been the case but Velázquez is much too clever: he is not content with providing an illusion, he must have an illusion of an illusion, because it is certain that he never used a mirror here.[3]

To start with, this picture is huge (317.5×276.8 cm) and a mirror to reflect the whole picture would have had to be at least some 2 m wide. Velázquez possessed a large number of mirrors but not one of them would have been large enough. The picture was finished in 1656 in accordance to Velázquez biographer Palomino,

---

[2]One of which might well be the painter himself (cf. p. XX).

[3]That Velázquez was interested in *trompe l'oeil* is well known: Kemp (1990) records that he had two Italian experts in *quadratura* brought to Madrid.

at which time no mirrors were easily available larger than about 90 cm in the longer dimension.[4] Although there are very many interpretations of *Las Meninas* (Foucault 1970; Searle 1980; Snyder 1980) I should like to venture some personal observations: gazing at a reproduction would never do the trick, as gazing at a photograph of the false vault at Sant'Ignazio in Rome by Andrea Pozzo (painted well after Velázquez' death) will never replace the illusion of a vault which the observer experiences when standing at the exact point marked by a plate on the church's pavement. The gaze of all the figures in *Las Meninas* that look to the front, the painter included, converge on a point a few metres in front of the picture, a point at which the relation between the observer and the picture becomes more intimate. It seems to me that it was at that point where the painter stood and painted most of the picture. All he then had to do was to paint his face over that of the model standing for him and to deal with the mirror at the back, which contains a very diffuse double portrait of Philip IV and his Queen Mariana.[5] This mirror is large by the standards of the day, perhaps some 90 cm high. The question is: how was this double portrait painted? Certainly, if the painter stood at the point suggested when doing the scene, it could not have been done because this is where the royal couple might have stood. But the canvas could have been moved to the position shown in the painting itself and then the double portrait added. Of course, given the properties of mirrors, the royal image would have appeared at double the length of the distance from them to the mirror, which probably explains why the artist painted them diffusely. The question of this double portrait and of the mirror on which it is supposedly reflected requires further discussion.

That the not visible one on which Velázquez appears to view the room shown in the painting never existed is now generally accepted. The important question is: what was him painting on the canvas that appears in the picture? Most scholars support the view that this was a double portrait of Phillip IV and his Queen Mariana. Kemp (1990) shows, in fact, by carefully analysing the perspective of *Las Meninas*, that such a portrait could have then been reflected on the back mirror. Carr et al. (2006) suggest on the other hand that what Velázquez is painting on the canvas shown could have been *Las Meninas* itself.[6] There are several reasons to support this view. If the painter wanted to state that he was painting the double portrait of the royal couple, why showing such a huge canvas? Such double portrait has never appeared in the careful inventories of the Alcázar. Moreover, it is perfectly well recorded that Philip hated his portrait being painted because he did not like to show the effects of age on his face (Stratton-Pruitt 2003, p. 139). At the time

---

[4]It was around 1680 when Colbert, having illegally spirited away some craftsmen from Venice for the manufacture of Faubourg Saint-Antoine, got them to make larger mirrors, one of which, 115 × 65 cm, was sold after his death for three times the value of a Rubens: a good mirror was worth in Venice as much as a whole warship.

[5]This mirror is indeed reminiscent of the one in the *Arnolfini*, which was then in the Alcázar Palace in which Velázquez worked.

[6]This is an opinion that has to be taken seriously, one of the collaborators in this book being Jaime Portús, the curator of El Prado expert on Velázquez.

Fig. 5.7 Francisco de Goya y
Lucientes: *Self-Portrait in the
Workshop*, 1790–1795.
Madrid, Museo de la Real
Academia de San Fernando

of this painting the king was 51, while the queen was only 22: if he was sensitive about his appearance *solo*, how much more would he had been next to a young girl in her prime. It is well known that Velázquez and the king were in a warm affectionate relation and the painter, aware of his master's qualms most likely would have proposed an ingenious solution: to introduce in the picture a fictitious mirror on which the royal couple would appear reflected and thus diffuse, disguising the effects of age. That such a hypothesis is plausible results from the fact that the mirror that appears in *Las Meninas* was never listed in the inventories of the Alcázar, despite the fact that it would have been far more valuable than the recorded pictures by Rubens that appear at the back.

We can now come back to important situations where mirrors are an essential tool for the artist, which is the case for self-portraits. Painters from Rembrandt to Arnold Schönberg, who painted dozens of them, most often avoided showing hands because the right-hand holding the brush would appear in the picture as a left hand. Not a problem, though, for Goya (Fig. 5.7), who must have painted, at least for a short time, with his left-hand, which under specular reflection appears to be the right one.

5.1.1.2 Right and Left in Visual Representations

Etchings or engravings, and tapestries, in the latter case because of the way the weavers work, entail specular reflection of the original design by the artist. This is, in principle, not a serious problem: William Blake, for instance, who produced numerous illustrated poems on copperplates, learnt to write his lines in mirror

Fig. 5.8 Raphael: *The Miraculous Draught of Fish, c.* 1515. *Left* tapestry, Rome, Vatican Museum. *Right* cartoon, London, Victoria and Albert Museum

writing so that they would appear normal when printed. Much more complex, though, is the situation with one the most famous suites of tapestries in existence. In 1515 Pope Leo X commissioned from Raphael a suite of ten tapestries for the Sistine chapel. Only seven of the original cartoons by Raphael are extant, all of them in the Victorian and Albert Museum in London. The surprising fact is that Raphael did not draw all his cartoons reversed, as discussed by Oppe (1944) amongst many others. In order to understand the effect of reversing a picture I show in Fig. 5.8 the tapestry of *The Miraculous Draught of Fish* with its cartoon on its right. Of course, in the tapestry Jesus points out to St Peter, correctly, with the right hand, but what is interesting is that it has been suggested (see Gaffron below) that, because of his position on the right, he should appear larger than in the inverted cartoon; the effect of the enhanced colour of his tunic in the tapestry, however, cannot be discounted. There are three cartoons that are not inverted. In the cartoon for *The Sacrifice at Lystra* (Fig. 5.9), for instance, the butcher is wielding his axe on the strength of his right arm and the implied movement is from left to right (to be discussed later). More importantly, Mercury in the cartoon is carrying the caduceus on his left hand, correctly in accordance to standard iconography so that it is wrong in the tapestry. An even clearer case is *The Conversion of the Proconsul* where an inscription in a plinth is not inverted in the cartoon as it should be, and in the cartoon of *The Healing of The Lame Man* Jesus points out to the latter with the right hand, rather than the left to allow for inversion.

Leaving the cartoons for a moment, we must discuss the relative significance of right and left in pictures. It is not sufficient to argue that left, *sinister* in Latin, was always considered not to be very much of a good thing, as the obvious evolution of the Latin word entails. Most people, on the other hand, would find the composition of the tapestry of *The Sacrifice at Lystra* somewhat awkward compared with its (correct) uninverted cartoon. No one would worry these days about the caduceus, but the movement of the butcher's axe from right to left appears somewhat unnatural (see Fig. 5.9). This is the type of problem, arising from inverted designs, that we must address. The reader must not expect, however, that I will provide a full answer to the questions we shall raise, because this problem is extremely complex; I shall try

however to review first the salient discussions in the literature before we can come to some plausible conclusions. It was Wölfflin (1928) who attempted a study of the problem of specular reflection in the design of paintings. He suggested that pictures are read from left to right, as script is. Arnheim (1954) developed this hypothesis and adduced that such direction of reading correlates with the domination of the right cerebral cortex (which controls the left-hand side field of vision).

The diagonal from bottom left to top right of a picture (which I shall call the first diagonal) is seen as going up, whereas the second diagonal, from top left to bottom right, is seen as descending. These views (remember that I am only reporting, not making any claims as to their validity) were reinforced by Gaffron (1950a,b, 1956), who claimed that there is a certain fixed path that observers follow normally within the picture space, which she calls the "glance curve", that moves from the left foreground to the right background (i.e., basically along the first diagonal). Because we look first at the left foreground, she argued, we tend to place ourselves in that position and to identify more readily with the figures on that position, whereas characters on the right appear more distant, psychologically as well as visually, for which reason we read them as larger and awesome. As a result, a figure placed on the left of a reversed cartoon, for instance, would appear larger on the tapestry, as already mentioned; Gaffron, in fact, claims that the right balance in Rembrandt's etchings must be observed in the original plates rather than in the reversed prints. The alleged relation between our left-to-right script as conditioning the reading of pictures from left to right has been used to suggest that, reading being less common in C16 or C17, explains why artists like Raphael or Rembrandt did not appear always to care to reverse their designs for cartoons or prints. Though plausible, this suggestion does not agree with the fact that even Picasso more than once signed a lithograph on the plate so that even his signature appeared reversed in the print.

The idea that script-reading habits determine the way in which observers read pictures, which appeared plausible at the time the work mentioned was published, has been disproved by modern eye-tracking work, although as we shall see, culturation is not insignificant and some of the proposals so far discussed still have an element of validity. The early work of Buswell (1935) shows that viewers' eyes follow short periods of *fixation* with rapid *saccadic* motions to other parts of the pictures. He found, for instance, that viewers of Seurat's *La Grande Jatte* fixated first on the people, irrespective of their position in the picture, rather than the background, and modern work on eye-tracking does not appear to show any distinctive initial preference for the left of the picture.[7] It must be remembered, however, that numerous factors may affect the result of eye-tracking studies. Hernandez Belver (1990), for instance, found differences between pupils of the Academy of Fine Arts and others, and Avrahami et al. (2004) observed gender differences in this respect. A very important result was found by McLaughlin and Kermisch (1997), namely that paintings containing cues suggesting left to right motion are preferred by dextrals over their mirror-reversed versions. This would account for my remark that the tapestry of *The Sacrifice at Lystra* (reflected as in the right of Fig. 5.9) appears less natural than the original cartoon.

A case where the subject of right and left in pictures is iconographically most significant is that of the *Annunciation*. It appears that since about C8 the iconographic convention, possibly of Byzantine origin, was to place the Virgin on the right field of the picture, which would agree with the views of Mercedes Graffon mentioned above. This convention, at least in Italy, appears to have been carefully maintained for many centuries. McManus showed, in fact, that in Berenson's eight-volume catalogue of Renaissance pictures, of 209 Annunciations 96.7 % show Gabriel entering from the left (McManus 1979, 2005). Pietro Cavallini's mosaic in the apse of S. Maria in Trastevere, c. 1295 is an example, as is also the almost contemporary one by Jacopo Torriti at S. Maria Maggiore. Hundreds of frescoes and paintings with this subject faithfully follow tradition but in illuminated manuscripts after 1415 Gabriel may come from the right. The first painting I know where this is the case is Dirck Bouts' *Annunciation* in Krakow (Fig. 5.10) from around the middle of C15 which I will discuss later, with others equally inverted from C16 on when the iconography becomes looser.

I shall now review recent work on the problem of right and left in pictures, after which we shall be able to see how this ideas stand in relation to the iconography of the Annunciation. We shall be concerned with western observers only, since it is within this culture that this iconography has arisen.

The first question I shall discuss is the perception of a *direction of motion* which of course can only be implicit in a picture, as we have already seen in the case

---

[7]Cf. Nodine and Krupinski (2004), Locher (2006); see also Locher et al. (2007), McLaughlin et al. (1982), McLaughlin (1986).

Fig. 5.10 Dirck Bouts:
*Annunciation, c.*1450.
Krakow, Princes Czartorisky
Museum

of *The Sacrifice at Lystra*. As already mentioned, it was found by McLaughlin and Kermisch (1997) that paintings with suggestions of left to right motion are preferred by dextrals over their mirror-reversed versions. The astonishing result is that Vallortigara (2006) found numerous and well-studied instances that show that the perception of left to right motion is also enhanced in the animal kingdom with respect to its reversed form. This appears to be related to the predominance of the right brain hemisphere (as mentioned before), and Vallortigara suggests that it is an evolutionary favourable trait for all the members of a given group to belong to the same dextral form, since variant subjects have advantages in fighting others, with consequential disruption of the group. Already Chatterjee (2001) had done careful experiments that show that normal subjects matched sentences they heard to pictures faster when pictures depicted the agent on the left and with the action proceeding from left-to-right. Later Chatterjee (2002) reported studies that show that subjects are likely to judge visual images more pleasing when any motion depicted in them is left to right, thus confirming the studies mentioned. Chatterjee went further and introduced the concept of *agency*. This may entail the representation of a physical action, such as pushing an object, or the conveying of a message, but could also be somewhat more abstract as for instance the agent being in a position of greater importance than another person related to it. Both the perception of implied motion and the agency effect are clearly related to the brain asymmetry, but cultural effects related to script direction are significant and superimpose themselves on the above. This was clearly demonstrated by Maass and Russo (2003) who studied Arabic subjects (more reliable for this purpose than Hebrew readers, since the latter are already used to right-to-left script in arithmetic). Chatterjee (2001) presents a very careful review of how left-to-right perception and agency are influenced by culturation. We can thus see that although Wölfflin and followers were not totally right, their defence of culturation cannot be totally disregarded. Dobel et al. (2007), in fact, produce evidence for the dominance of the latter in this context, as do Christman and Pinger (1997) and Chokron and De Agostini (2000). For my purposes, I shall consider *direction of motion* as part of the *agency* effect.

The second problem that I want to discuss is the *power of the first diagonal*, again originating from the old work of Gaffron but that nevertheless very recent work suggests it is significant. Pérez Gonzalez (2007) made a careful study of photographic work in the nineteenth century, when families were large and family portraits very popular. She found that western portraits posed the families starting from the youngest on the left and ascending to the tallest on the right, that is following the first diagonal. Iranians, who as Farsi speakers write from right to left, consistently used the opposite convention. We again have a rule largely valid for western subjects (the power of the first diagonal for composition purposes) but subject to culturation.

The third and last important effect that we must discuss is the fact that in portraits the apparent right–left symmetry of the cheeks is violated. McManus and Humphrey (1973) examined 1474 portraits and showed that 60 % show the left cheek, and that this proportion is even higher for portraits of women, 68 %, as also discussed by McManus (2005). Suitner and Maass (2007) found that gender differences disappear after 1848, following the greater status of women in society. Chatterjee (2011) shows that the bias to depict the left cheek of women decreased from the C15 to C20. For women, the ratio of left-to-right cheek depictions was about 8 to 1 in C15, and diminished gradually to a ratio of slightly above 1 to 1 in C20. Grüsser et al. (1988) and Suitner and Maass (2007) examined a new set of portraits to test the agency hypothesis in portrait profiles. They replicated the original observations of a general bias for artists to paint portraits depicting more prominently the left cheek with a greater frequency than the right. However, they also found that this bias depended on the gender of the artist. The findings obtained on male artists did not generalize to women. Female artists, unlike male artists, were not likely to portray women sitters with a left orientation rather than with a right orientation. The authors suggest that male artists prefer to select a spatial orientation for their female sitters that reflects their stereotypical view of females as passive, thus showing their left cheeks, whereas women artists appear less subject to these stereotypic views. ten Cate (2002) studied portraits of their professors in six German and two Dutch universities where during several centuries such portraits were collected. Before 1600, 90 % of them showed the right cheek of the sitters, a polarization that did not last after C18, although right-cheeked originals were perceived as more scientific than left-cheeked ones. The prevalence of women portraits showing the left cheek is supposed to be associated with the fact that this is the side of the woman's face that best expresses emotions, under the control of the right brain hemisphere.

Fig. 5.11 *Left* Domenico Veneziano: *Annunciation, c.* 1442–1448. Cambridge, Fitzwilliam Museum. *Right* its "mirror reflection" (Courtesy of Professor McManus)

I shall now try to show how the three effects discussed account for the canonical iconography of the Annunciation until some time in C15. I shall consider for this purpose just one example, which I owe to Chris MacManus and which is shown in Fig. 5.11. On the left we show the original picture and on the right its mirror reflection. There is, of course, a disturbing change of perspective but the original on the left appears more natural. This is probably the result of a combination of the three effects discussed above. The *agency* effect of Gabriel is clear, first as arriving into the scene, thus from the left, and also because of his role as messenger, emphasized by his raised right hand. Because the archangel is kneeling and the Virgin is standing the composition respects the *power of the diagonal* and, finally, because of her position, the Virgin shows her *left cheek*, as preferred for women.

In order to assess how far the analysis above may be significant it is useful to consider what happened when painters decided to produce more expressionistic depictions of the Annunciation, some time around C15. I shall discuss for this purpose two such interpretations due respectively to Dirck Bouts, middle of C15 (Fig. 5.10) and Lorenzo Lotto, *c.* 1527 (Fig. 5.12). Despite the enormous difference in size (they are about 25–30 and 166 cm in height, respectively), the composition is remarkably similar. In order to achieve the desired expression, the Virgin is presented in both pictures kneeling in an orant position, full face to the front, thus not requiring a choice of presentation of either cheek. Because of this crouching position if the Virgin had been on the right in the Bouts it would have broken the first diagonal rule, which the artist preferred to keep. Although the Virgin and Gabriel reach the same height in the Lotto, the strong figure of the Father emphasizes the first diagonal, which is accentuated by the similarity of the strong colour of that figure and that of the Virgin. It appears that when the "mood" of the picture requires an inversion of the positions of the two main figures the artist still tries to keep the power of the first diagonal. In the inverted *Annunciation* of Andrea del Sarto (1512–1513) at the Galleria Palatina, Palazzo Pitti, Florence, although Virgin and angel are about the same height, the latter is supported at his back by three tall figures that enhance the diagonal effect. Perhaps the most prolific painter of Annunciations is El Greco (1541–1614) of whom at least some ten examples are known, all of them inverted. El Greco's angels are never kneeling but stretched to their full height and are either considerably taller than the Virgin or floating at a good height above her.

Fig. 5.12 Lorenzo Lotto:
Annunciation, c. 1527.
Recanati, Pinacoteca
Comunale

Thus, again, for compositional purposes the artist has respected the value of the first diagonal.

A word of caution is required about the above discussion, since the weight of tradition must not be discarded. One of the major determinants of composition, I suggested, is the power of the first diagonal, for which some experimental evidence exists. Nevertheless, when such an effect is unlikely to be significant, the left position of the archangel is normally respected, as when this figure and that of the Virgin are displayed in adjoining spandrels. A good example is Masolino's *Annunciation* in the Cappella di Santa Caterina at San Clemente, Rome, early C15.

It is interesting that the standard iconography is not necessarily maintained when reliefs or sculptures are considered, a fact that unfortunately is not discussed in the literature, although these objects were produced during the period when that iconography was most faithfully followed. Figures 5.13 and 5.14 show the Annunciations by Bonanno Pisano (1180) and Andrea della Robbia (1475), respectively. In Rome, on a portal on the Muro della Suburra in via Tor de' Conti, adjoining the Casa dei Cavalieri di Rodi, there is a good example of such a relief in the tympanum, probably C16–C17. On the other hand, the high relief in Florence by Donatello is orthodox. But there are many other inverted examples, of which perhaps the most remarkable is the work by Veit Stoss in the polychrome wood carvings of the magnificent altarpiece at the Basilica of St. Mary in Krakow, made in the period 1477–1489. Unfortunately, I am not aware of any explanation why this difference

Fig. 5.13 Bonanno Pisano:
Annunciation, 1180 (bronze).
Pisa, Ranieri Portal of Duomo

Fig. 5.14 Andrea della
Robbia: *Annunciation*, 1475
(ceramic). Chiusi, Santuario
della Verna

between carvings and paintings exists, although it is possible that carvers, given the
nature of their medium, do not have a rule for the first diagonal, which appears to
be of some importance for painters.

5.2 Rotations, Translations, and the Inversion

Reflections are the fundamental symmetry operations, because any rotation can be
expressed as the succession of two reflections. In Fig. 5.15 we prove, in fact, the
following result:

- *The succession of two reflections on planes that form an angle α is a rotation by
  twice this angle about the line of intersection of the two planes.*

Fig. 5.15 Two reflections on planes separated by an angle α are equivalent to a rotation by 2α

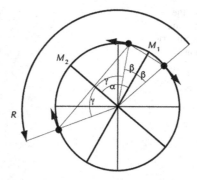

$\beta + \gamma = \alpha$ The angle of the rotation R is $2\beta + 2\gamma = 2\alpha$

Fig. 5.16 A sixfold rotation ornament in brick in the outside wall of the Basilica del Santo Sepolcro, San Stefano, Bologna, C12

Despite this result, rotations have been much used independently in decorating pots, but I illustrate an early use as a mural decoration in brick in Fig. 5.16 (notice that the $\pi/3$ pattern is not very accurate). In the C14, Moorish craftsmen at the Alhambra combined rotations with their repetitions by translation to obtain the most varied and complete set of two-dimensional symmetries until then known (Figs. 5.17 and 5.18).

Translations and rotations gave origin to mathematical objects that have an enormous importance in physics but which are also most interesting because of their, in certain cases, astonishing symmetry properties. They are called *vectors* but the discussion I shall give is the result of more than half a century of mathematical controversy (Altmann 1992). If we consider a segment of a straight line as the limiting concept of a rod, when you allow its diameter to become infinitesimally small, that is, as small as we wish, then a *vector* is a *directed* rod of infinitesimal diameter, that is a rod the two ends of which are intrinsically distinct (*intrinsically* meaning that they must not be distinguished by any marks we might add). Now, there are two ways in which the two ends of a rod may become intrinsically distinct. The simplest is to impress a velocity to the rod (Fig. 5.18a), whereby one end

Fig. 5.17 A tile from the
Alhambra, (C 14) showing
fourfold rotational symmetry
repeated by translation

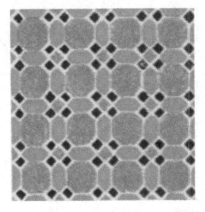

Fig. 5.18 (a) A polar vector;
(b) an axial vector

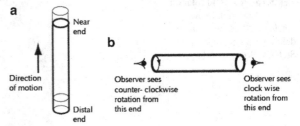

becomes the near end, that is the end which is nearer the point towards which the rod
is moving, and the other its distal end. But the two ends become also automatically
distinct if we impress a rotation of the rod, because (see Fig. 5.18b) from one end an
observer will see the rotation counter-clockwise and from the other it will be seen
clockwise.

So, we have two types of vectors, velocity-like or *polar vectors* and rotation-
like or *axial vectors*, and I am afraid I have to reveal a weakness of mathematical
notation: both objects are represented traditionally by the same identical symbol,
which is an arrow. For polar vectors the arrow is pretty obvious, since it is placed
in the direction in which the vector advances (Fig. 5.19a). In the case of axial
vectors the head of the arrow is conventionally placed at the end from which
the rotation is seen counter-clockwise, as shown in Fig. 5.19b,c. Polar and axial
vectors have entirely different symmetry properties, and this is the reason why
I discuss them here. I must first stress that it is no good to try to obtain such
transformation rules by depicting the representative arrows only, which produces
no more than extremely interesting mistakes: it is necessary to take careful account
of the underlying definition.

In Fig. 5.20, we transform a polar vector under a reflection on a plane parallel
to the vector. Of course, given the properties of mirrors, if the rod moves up in the
object field, it also moves up in the image field. Thus, in both fields, the vector
and its transform are represented with parallel arrows, as one would expect. This
is not the case for axial vectors, which is responsible for the most delicious falsely

Fig. 5.19 Representation of polar (a) and axial (b, c) vectors by *arrows*

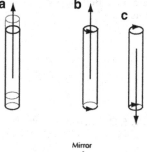

Fig. 5.20 Transformation of polar vectors under reflection

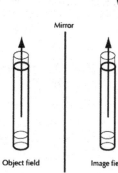

Fig. 5.21 Transformation of axial vectors under reflection

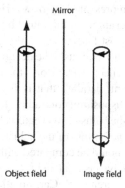

so-called "paradoxes". In Fig. 5.21 we consider the reflection of an axial vector on a mirror parallel to it. Because of the transformation properties under reflection, if the rod is rotating counter-clockwise in the object field its image rotates clockwise in the image field. In the object field, therefore, the representative arrow must be placed upwards because from its head we see the rotation counter-clockwise, but in the image field it must be placed downwards because it is from below that we see the image rod so rotating. If, as often done in physics, we use the arrows without the underlying rotating rods, we have the counter-intuitive result that they flip over under reflection on a plane parallel to them.

It is interesting to mention that the gravitational force, as one would expect, is a polar vector, so that it behaves decently under reflection, but that the electromagnetic forces are axial vectors and thus one has to be careful in transforming their

Fig. 5.22 Borromini's
balustrade

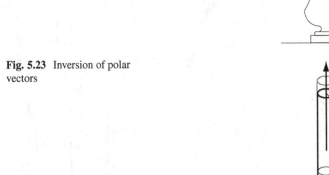

Fig. 5.23 Inversion of polar
vectors

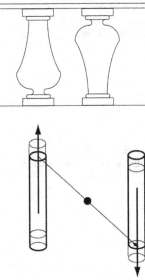

The black circle is the centre of inversion

representative arrows. How is this reflected in art? I am afraid that I know only one
example. It might not be an exaggeration to say that until Antoni Gaudi the architect
most fascinated by the gravitational force was Francesco Borromini (1599–1667)
and, curiously, he designed some balustrades where the balusters, that are normally
reflected from one to the next as polar vectors are, that is, parallel, are instead
anti-parallel, that is flipped over in the manner that axial vectors do. Thus their
disposition contradicts the gravitational force, not just theoretically when imagining
the balusters as vectors, but practically, since in general the heavier mass is placed at
the bottom of the baluster shafts and not at their top as illustrated in Fig. 5.22, which
should be compared with the reflection of axial vectors in Fig. 5.21. Borromini must
have been taken by this conceit since he deployed it in two buildings in Rome. The
church of S. Carlino alle Quattro Fontane, on the street of that name, was built in
the period 1634–1641, and uses this device extensively in the cloister, the first part
of the church that he built, so that this is probably the oldest of the two examples.
The second is the Oratorio dei Filippini (1637–1667), in Piazza della Chiesa Nuova,
where the balusters appear in a less extensive way, on the balcony at the centre of
the main elevation. That Borromini knew that he was imitating the behaviour of the
electromagnetic forces, rather than foe gravity, is not unlikely: it is impossible. Yet,
the inventions of a great artist are sometimes amazing.

It will be useful to say a few words about the inversion I, which is an important
symmetry operation because it entails the simplest possible symmetry element: a
mathematical point, called the centre of inversion. How it works will be obvious
from Fig. 5.23, where we invert a polar vector, which is indeed inverted as one
would expect. This is not the case with polar vectors, which are invariant under
inversion, as shown in Fig. 5.24. (Remember that the head of the arrow is always
placed at the end from which the rotation is seen counter-clockwise.)

Fig. 5.24 Axial vectors are
invariant under inversion

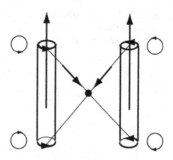

The black centre is the centre of inversion.
The small circles show the rotation sense
when viewed from *outside* the rod.

5.3   Vectors and Spinors

Vectors, whether polar or axial, can be expressed as triples of real numbers in terms of some unit vectors along three non-coplanar and non-collinear directions, as illustrated in Fig. 5.25. By a standard rule of addition of vectors, the vector **a** of the figure may be written as follows,

$$a = a_1\mathbf{e}_1 + a_2\mathbf{e}_2 + a_3\mathbf{e}_3 = (a_1, a_2, a_3),$$

where the bracket is a convenient shorthand for the vector whenever the underlying unit vectors may be taken as given.

With this notation the effect of the inversion I on vectors can be written as follows:

$$I(a_1, a_2, a_3) = (-a_1, -a_2, -a_3), \quad \text{(polar vector)}.$$

$$I(a_1, a_2, a_3) = (a_1, a_2, a_3), \quad \text{(axial vector)}.$$

For a triple of real numbers to qualify as a vector not only one of the two properties above must obtain, but they must also transform in a specific way under rotations, although these transformation rules are more involved and I shall take them for granted.

When we look at Fig. 5.25 we may conclude that polar vectors span the ordinary three-dimensional space, in the sense that every point of the latter may be denoted by a unique vector, once the coordinate axes are chosen. The world, alas, is rather more interesting than that, as the French mathematician Élie Cartan (1913) discovered. Cartan used a multiplication rule for triples that is called the *tensor product* and adapted it to doublets $(\lambda\mu)$ that he defined, called *spinors*, so that

$$(\lambda_1\mu_1) \times (\lambda_2\mu_2) = (\lambda_1\lambda_2, \lambda_1\mu_2, \mu_1\lambda_2, \mu_1\mu_2).$$

Fig. 5.25 Vectors expressed
in terms of three
non-collinear unit vectors

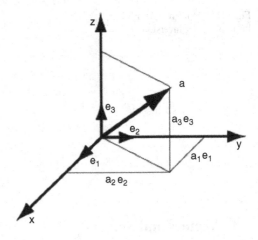

The next ingenious step is to re-write the above expression:

$$(\lambda_1\lambda_2, \lambda_1\mu_2, \mu_1\lambda_2, \mu_1\mu_2) : (\lambda_1\lambda_2, \frac{1}{2}(\lambda_1\mu_2 + \mu_1\lambda_2), \mu_1\mu_2) + \frac{1}{2}(\lambda_1\mu_2 - \mu_1\lambda_2).$$

The idea here is that the triplet on the right-hand side is symmetrical under exchange of 1 and 2, whereas the singlet (last term) is anti-symmetric. It is a general mathematical principle that such quantities do not "mix" and must be treated separately. If we define, with Cartan, (after all, *he* invented $(\lambda\mu)$),

$$I\lambda = \lambda, I\mu = \mu, \quad \text{for all } \lambda, \mu,$$

then the triplet behaves like an axial vector under inversion (compare with the corresponding equation above) and with a suitable choice of the transformation rule for $(\lambda\mu)$ under rotations it can be shown that it is indeed an axial vector. However, we could also define a different type of doublet $(\lambda\mu)$ with the following transformation property under inversion:

$$I\lambda = \mathbf{i}\lambda, I\mu = \mathbf{i}\mu, \quad \text{for all } \lambda, \mu, \text{ with } \mathbf{i} \text{ imaginary units}, \mathbf{i} = \sqrt{-1}, \mathbf{i}^2 = -1.$$

In this case each term of the triplet above is multiplied by $\mathbf{i}^2 = -1$ when acted upon by the inversion I, whence the triplet itself behaves like a polar vector under inversion. This shows that polar vectors, that we take to denote all points of space (by which we mean the everyday' space) are themselves manifestations of more basic objects, the spinors.[8] It was not Cartan who called these remarkable doublets *spinors*: some 10 years after his work they were independently re-discovered by

---

[8] For the sake of the interested reader: the singlet above, in either case, is some sort of a scalar, scalars being just real numbers.

Pauli who gave them this name because he found that they exactly describe the properties of some, until then unknown, coordinate of the electrons, that represents their *spin*. Thus, they become essential tools in the study of quantum mechanics.

5.4 Some Properties of Rotations

We must now understand some strange properties of rotations that will help us later grasp some ideas about *quaternions*. The question is that some continuity conditions must be satisfied, which must properly be studied by *topology*, but I cannot get here into the full panoply of concepts required by this science. I shall try instead to use some plausible arguments to give the reader some idea of the problem. Before we get into this it will be useful to realize that whereas the product ab of two numbers commutes, that is, equals ba, this property is not obeyed by rotations. We show in fact in Fig. 5.26 that the succession of two rotations (also called their *product*) depends on the order in which the rotations are performed.

We can now consider a major problem, which concerns the possibility of ambiguity for some specific rotations, as illustrated in Fig. 5.27. On the left side a of this figure, R_1 and R_2 are both rotations by angles of the same magnitude, but conventionally R_1 and R_2 are said to be positive (counterclockwise) and negative (clockwise), respectively. Clearly, an ambiguity arises when we allow the angle in both cases to reach π, since now R_1 and R_2 become identical (i.e., they transform P into the same identical point), as shown in b. In order to resolve the ambiguity we decide to take, as we have done for the two rotations in a,

$$R_2 = (-1)R_1.$$

In c we extend this result whenever R_1 and R_2 are in fact identical rotations but in opposite senses. We must now face a remarkable result that concerns the identity rotation E, the rotation by nothing, which therefore changes nothing. If in c we allow R_2 to become such a rotation, that is the identity E, R_1 will become a rotation by 2π, which by continuity from the above equation will give us that the rotation by 2π will *not* be the identity but rather its negative:

$$R(2\pi) = (-1)E, \quad \text{that is:} \quad R(2\pi)^2 = R(4\pi) = (-1)^2 E^2 = E.$$

If the reader is surprised by this result he or she is in good company since no one could even imagine it until it was discovered by Élie Cartan in 1913, after which time very careful topological analysis showed its correctness beyond any possible doubt. It helps to be reconciled with this result, however, to consider the Möbius strip, illustrated in Fig. 5.28. If you imagine that you mark a point at the top edge of the figure, and then draw a continuous line, once you have moved around a whole circle, that is 2π, you do not return to that point but only underneath it. You require another turn of 2π to return to the original point. That is, the identity "rotation" on the strip is 4π.

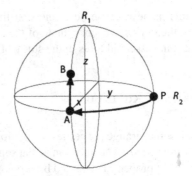

R_1 is a clockwise rotation around z by 90°

R_2 is a clockwise rotation around z by 45°

A is the result of R_2 (which leaves the starting point P
            invariant) followed by R_1

B is the result of R_1 followed by R_2

Fig. 5.26 Rotations do not necessarily commute

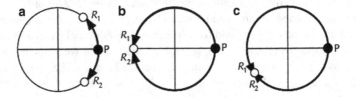

Fig. 5.27 Ambiguities in the definition of some rotations

Fig. 5.28 The Möbius strip

5.5 Hamilton, Quaternions, and Rotations

Sir William Rowan Hamilton (1805–1865), Irish child prodigy, Astronomer Royal of Ireland at 21, knighted at 30, named by the American National Academy of Sciences as the greatest living scientist in 1865, creator of some of the most influential modern ideas in mathematical physics, is thus better qualified than most to be judged by La Rochefoucauld's *Maxime* 190: "*Il n'appartient qu'aux grands homes d'avoir de grands défauts.*" Hamilton had attempted for a long time to extend

Fig. 5.29 The rotation by $\pi/2$, R, repeated twice changes the sign of a vector

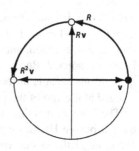

the concept of the imaginary unit $i = \sqrt{-1}$, $\mathbf{i}^2 = -1$. He visualized this property by representing the imaginary unit as a $\pi/2$ rotation, because as shown in Fig. 5.29, the square of such rotation changes the sign of a vector. Of course, he could not possibly have been aware of the argument discussed in the last section, and he could not be blamed for that. But one can blame him for his extraordinary stubbornness in rejecting the evidence of his own results, which clearly contradicted his view of the imaginary unit. His aim was to extend the concept of imaginary numbers and he tried to achieve this at first by introducing one extra imaginary unit. Then, on Monday 16 October 1843 (see Hamilton 1844), he hit on the fact that three in total where required:

$$\mathbf{i}, \mathbf{j}, \mathbf{k}, \quad \text{with} \quad \mathbf{i}^2 = \mathbf{j}^2 = \mathbf{k}^2 = -1.$$

His stroke of genius in creating these new units was to realize that in order to form a decent algebra they could not commute (as indeed we have seen is the case for rotations):

$$\mathbf{ij} = \mathbf{k}, \quad \mathbf{ji} = -\mathbf{k}; \quad \mathbf{jk} = \mathbf{i}, \quad \mathbf{kj} = -\mathbf{i}; \quad \mathbf{ki} = \mathbf{j}, \quad \mathbf{ik} = -\mathbf{j}.$$

Hamilton could then verify that these new objects satisfy all the operational rules of ordinary algebra, except for the commutation rule. He then went further and defined a *quaternion* in terms of a scalar and an object for which he in this context invented the word *vector*:

$$\mathbf{A} = [a, A_1\mathbf{i} + A_2\mathbf{j} + A_3\mathbf{k}] = [a, \mathbf{A}].$$

The problem arose when Hamilton defined a vector as a "pure" quaternion, (for which the scalar part vanishes) making no distinction between such a quaternion and its vector:

$$\mathbf{A} = [0, \mathbf{A}].$$

Thus, he would write

$$\mathbf{i} = [0, \mathbf{i}],$$

and identify it with a rotation by $\pi/2$, for the reasons explained above. Notice that then $\mathbf{i}^4 = (-1)^2 = 1$. This was all right for Hamilton who visualized $\mathbf{i}^4$ as four times a $\pi/2$ rotation, and thus as the identity but, as we have seen, this result was later contradicted. Because of this, the whole question of how to handle rotations by quaternions became mired in confusion from which the theory of vectors had to be rescued at the end of the century.

The irony of it is that the whole subject of rotations had already been beautifully treated before Hamilton by an obscure French banker, Olinde Rodrigues (1794–1851) in 1840. In our vector and quaternion notation, which he did not use, he identified a rotation by α around an axis denoted by the vector $\mathbf{v}$, $R(\alpha, \mathbf{v})$, as follows:

$$R(\alpha, \mathbf{v}) = [f(\alpha), g(\alpha)\mathbf{v}],$$

where $f(\alpha)$ and $g(\alpha)$ are simple (trigonometric) functions. The important result was that Rodrigues was then able to multiply two such objects and thus obtain the angle and axis of the resulting rotation. Although Hamilton became aware of such formula through the work of Cayley, he never paid attention to it because it did not fit his preconceived ideas. The problem was that Rodrigues, in doing this work, pragmatically ignored that it entailed as an identity a rotation by 4π, whereas the rotation by 2π implied a factor of -1, a result that Hamilton could not explain to himself and thus could not countenance. It is not improbable that this internal contradiction was partly responsible for the misery of the last 20 years of Hamilton's life.

The formulae for multiplication of rotations discovered by Rodrigues are now much used in such applications as robot controls, satellites motion, tracking of eye movements, and so on. In art they are at the core of one of the more lively contemporary forms of applied art: computer animations, the first such application being for the game called *The Tomb Raider*, where the ineffable Lara Croft was able to move courtesy of Olinde Rodrigues. And, of course, they are of crucial importance in quantum mechanics where rotations appear in a variety of phenomena such as nuclear structures, the rotational spectra of molecules and properties of crystal structures.

5.6 Quaternions *vs* Clifford Algebras

We have seen that Hamilton found it necessary, in order to construct an algebra with more than one imaginary unit (the square of which must be -1), to introduce three such units, that anti-commute:

$$\mathbf{i}, \mathbf{j}, \mathbf{k}, \quad \text{with} \quad \mathbf{i}^2 = \mathbf{j}^2 = \mathbf{k}^2 = -1, \quad \text{with} \quad \mathbf{ij} = -\mathbf{ji} = \mathbf{k}.$$

Hamilton identified the quaternion units with quadrantal rotations (rotations by $\pi/2$) but this is wrong as we have seen: although counter-intuitive for him, the

Fig. 5.30 The products $\sigma_x\sigma_y$ and $\sigma_y\sigma_x$

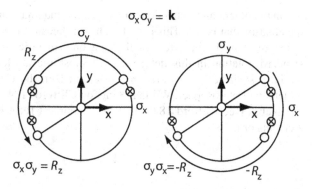

quaternion units are binary rotations (rotations by π). This comes from the fact that their square, a rotation by 2π, is *not* the identity that can be associated with $+1$, as assumed by Hamilton, but rather the anti-identity associated with -1, so that the true identity, its square, (rotation by 4π) is $+1$. It is because of this reason that the square of a quaternion unit is -1. The real problem here is that, as we have discussed, rotations entail somewhat complex topological problems. Moreover, although the algebra of quaternions is the algebra of rotations, Frobenius proved in 1878 that no further extensions to a higher number of imaginary units (higher dimensions than 3) are possible. So, rotations remain an artefact of three-dimensional space and thus lose generality.

The problem that we now want to consider is whether the imaginary units such as **k** are themselves products of other more elementary units. In order to answer this question we must remember that **k** is a binary rotation around the z axis. We can now appeal to the result in italics at the top of Sect. 5.2, which I now transcribe:

- *The succession of two reflections on planes that form an angle α is a rotation by twice this angle about the line of intersection of the two planes.*

This means that if σ_x and σ_y are reflection planes perpendicular to each other, and thus separated by $\pi/2$, then their product is a binary rotation around the z axis. The product $\sigma_x\sigma_y$ therefore equals **k**, a binary rotation about the z axis. One must be careful with this product because these operations appear to commute, unless one remembers that the two binaries that appear as a result of the two products may be one the negative of the other. To demonstrate that this is so requires a bit of topology, but I hope that the reader will find the treatment of Fig. 5.30 plausible.

In order to understand the figure it must be appreciated that the usual convention is that in a product such as $\sigma_x\sigma_y$, the first operation to be performed is the one on the right. Thus the motif to be transformed in the first quadrant, made up of two little circles, must first be reflected on the vertical plane and goes to the second quadrant. It is because of this that we take the path of R_z to go through that quadrant and thus R_z to be positive. Likewise, the product $\sigma_y\sigma_x$ on the right must be taken to be negative.

On the other hand, reflections are no longer imaginary units because σ^2 is always the identity, that is $+1$. This is pretty obvious because in the product σ^2 the second reflection cancels the effect of the first one. In order to adapt ourselves to the standard notation in this field, we shall substitute the letter e for the σ, so that we have so far $e^2 = 1$, $e_1 e_2 = -e_2 e_1$. In fact, the Cambridge mathematician, athlete, and philosopher, William Kingdon Clifford (1845–1879) invented in 1878 an algebra (see Clifford 1878) that in its simplest form is given in terms of four elements or units,

$$1, e_1, e_2, e_3,$$

with the following properties:

$$e_1^2 = e_2^2 = e_3^2 = 1, \quad e_1 e_2 = -e_2 e_1, \quad \text{and so on.}$$

The first set of conditions agrees with the square of a reflection σ shown above and the anti-commutation condition matches the same property for reflections, as already mentioned.

It is clear that the algebra so defined is more straightforward than the quaternion algebra, avoiding the problems of the duality of the binary rotations, and also that it contains the quaternion algebra as a special case. That this is so can be seen as follows: if we take e_1 and e_2 to be σ_x and σ_y respectively, then from Fig. 5.30, $e_1 e_2 = \sigma_x \sigma_y = \mathbf{k}$, so, as averred before, the quaternion units are just product of the Clifford units. Moreover, the Clifford algebra can easily be extended for any number of basic units, which therefore facilitates the study of symmetries in spaces of dimension larger than three, which has important applications in quantum mechanics. And do not forget that we started with mirrors with the Venus of Willendorf and we end with them as the fundamental operations. Yes, all is done with mirrors!

References

Altmann, S. L. (1992). *Icons and symmetries*. Oxford: Clarendon Press.

Altmann, S. L. (2002). *Is nature supernatural?: A philosophical exploration of science and nature*. Amherst, NY: Prometheus Books.

Arnheim, R. (1954). *Art and visual perception: A psychology of the creative art*. Berkeley: University of California Press.

Avrahami, J., Argaman, T., & Weiss-Chasum, D. (2004). The mysteries of the diagonal: Gender-related perceptual asymmetries. *Perception & Psychophysics, 66*, 1405–1417.

Buswell, G. T. (1935). *How people look at pictures: A study of the psychology of perception in art*. Chicago: Chicago University Press.

Carr, D. W., Bray, X., Elliottt, J. H., and Portús, J. (2006). *Velázquez*. London: National Gallery.

Cartan, É. (1913). Les groupes projectifs qui ne laissent invariante aucune multiplicité plane. *Bulletin de la Société Mathématique de France, 41*, 53–96.

Chatterjee, A. (2001). Language and space: Some interactions. *Trends in Cognitive Science, 5*, 55–61.

Chatterjee, A. (2002). Portrait profiles and the notion of agency. *Empirical Studies of the Arts, 20(1),* 33–41.

Chatterjee, A. (2011). Directional asymmetries in cognition: What is left to write about? In A. Maass & T. W. Schubert (Eds.), *Spacial dimensions in social thought* (pp. 189–210). Berlin: De Gruiter.

Chokron, S., & De Agostini, M. (2000). Reading habits influence aesthetic preference. *Cognitive Brain Research, 10,* 45–49

Christman, S., & Pinger, K. (1997). Lateral biases in aesthetic preferences: Pictorial dimensions and neural mechanisms. *Laterality, 2,* 155–175.

Clifford, K. (1878). Applications of Grassmann's extensive algebra. *American Journal of Mathematics Pure and Applied, 1,* 350–358.

Dobel, C. G., Diesendruck, G., and Bölte, J. (2007). Age influence spatial representations of actions: A developmental, cross-linguistic study. *Psychological Science, 18(6),* 487–491.

Foucault, M. (1970). Las Meninas. In *The order of things: An archaeology of the human sciences.* New York: Vintage Books.

Gaffron, M. (1950a). Right and left in pictures. *Art Quarterly, 13,* 312–331.

Gaffron, M. (1950b). *Die Radierungen Rembrandts. Originale und Drucke: Studien über Inhalt und Komposition.* Mainz: Kupferberg.

Gaffron, M. (1956). Some new dimensions in the phenomenal analysis of visual experience. *Journal of Personality, 24,* 285–307.

Gregory, R. (1997). Mirrors. In *Mind.* Oxford: W. H. Freeman.

Grüsser, O.-J., Selke, T., & Zynda, B. (1988). Cerebral lateralization and some implications for art, aesthetic perception and artistic creativity. In I. Rentschler, B. Herzberger, & D. Epstein (Eds.), *Beauty and the brain: Biological aspects of aesthetics* (pp. 257–293). Boston, MA: Birkhauser Verlag.

Hamilton, W. R. (1844). On quaternions, or on a new system of imaginaries in algebra. *Philosophical Magazine, 25(3),* 489–495.

Hernandez Belver, M. (1990). La experiencia art'stica y el lado derecho del cerebro. *Arte, Individuo y Sociedad, 3,* 99–109.

Kemp, M. (1990). *The science of art: Optical themes in Western art from Brunelleschi to Seurat.* New Haven: Yale University Press.

Locher, P. J. (2006). The usefulness of eye movement recording to subject an aesthetic episode within visual art to empirical scrutiny. *Psychology Science, 48,* 106–114.

Locher, P., Krupinsky, E. A., Mello-Thoms, C., and Nadine, C. F. (2007). Visual interest in pictorial art during an aesthetic experience. *Spatial Vision, 21(1–2),* 55–77.

Maass, A., & Russo, A. (2003). Directional bias in the mental representation of spatial events: Nature or culture? *Psychological Science, 14,* 296–301.

McLaughlin, J. P. (1986). Aesthetic preference and lateral preferences. *Neuropsychologia, 24,* 587–590.

McLaughlin, J. P., Dean, P., & Stanley, P. (1982). Aesthetic preference in dextrals and sinistrals. *Neuropsychologia, 21,* 147–153.

McLaughlin, J. P., & Kermisch, J. (1997). Salience of compositional cues and the order of presentation in the picture reversal effect. *Empirical Studies in the Arts, 15(1),* 21–27.

McManus, I. C. (1979). *Determinants of laterality in man.* Ph.D., Cambridge.

McManus, I. C. (2005). Symmetry and asymmetry in aesthetics and the arts. *European Review, 13(Supp. No. 2),* 157–180.

McManus, I. C., & Humphrey, N. (1973). Turning the left cheek. *Nature, 243,* 271–272.

Nodine, C., & Krupinski, E. (2004). How do viewers look at artworks? *Bulletin for Psychology and the Arts, 4,* 65–68.

Oppe, A. P. (1944). Right and left in Raphael's cartoons. *Journal of the Warburg and Courtald's Institutes, 7,* 82–94.

Pérez Gonzalez, C. (2007). Defining a model of representation for 19th century Iranian Portrait Photography. *Photoresearcher, 8(10),* 17–22.

Rodrigues, O. (1840). Des lois géométriques qui régissent les déplacements d'un sistème solide dans l'espace, et de la variation des coordonnées provenant de ses déplacements considérés indépendamment des causes qui peuvent les produires. *Journal de Mathématiques Pures et Appliquées, 5,* 380–440.

Searle, J. R. (1980, Spring). Las Meninas and the Paradoxes of pictorial representation. *Critical Inquiry, 6,* 477–488.

Snyder, J. C. (1980, Winter). Las Meninas and the Paradoxes of pictorial representation. *Critical Inquiry, 6,* 429–447.

Stratton-Pruitt, S. (2003). Velázquez's Las Meninas: An interpretive primer. In S. L. Stratton-Pruitt (Ed.), *Velázquez's Las Meninas* (pp. 124–140). Cambridge: Cambridge University Press.

Suitner, C., & Maass, A. (2007). Positioning bias in portraits and self-portraits: Do female artists make different choices? *Empirical Studies of the Arts, 25(1),* 71–95.

ten Cate, C. (2002). Posing as professor: Laterality in posing orientation for portraits of scientists. *Journal of Nonverbal Behaviour, 26(3),* 175–192.

Vallortigara, G. (2006). The evolutionary psychology of left and right: Costs and benefits of lateralization. *Developmental Psychobiology, 48,* 418–427.

Wölfflin, H. (1928). Über das rechts und links im Bilde. In H. Wölfflin (Ed.), *Gedanken zur Kunstgeschichte: Gedrucktes und Ungedrucktes* (3rd ed., pp. 82–90). Basel: B. Schwabe.

Part II
The Complex Route

The essays collected in Part I trace the origins of linear perspective in the Renaissance culture and explore its impact from painting to mathematical thought. Like linear perspective, complex numbers and probability originated in the Renaissance. Is this just a coincidence? Is there any meaningful link between linear perspective, which gave painting a new dimension, and complex numbers, which seem to be doing the same to physics? Part II is focused on exploring this link and helping the reader see the "art" involved in connecting complex numbers and probability through the notion of "quantum probability amplitude."

To start with, it may be worth reconsidering two meanings of "symmetry" that remained separated in the mathematic and artistic culture of the classic age: one related to the notion of "measure" and the other to the notion of "correct proportion." For ancient mathematicians, symmetry (*sun métron*) meant commensurability between two quantities. Artists merged this mathematical, static, and abstract notion with another form of symmetry—"aesthetic," dynamic, and relational—introduced by Polycletus and adopted as an operator of the architectural theory by Vitruvius (first century BC). The Renaissance notion of *symmetry*—at the same time mathematic and aesthetic, abstract and dynamic—captures the supreme principle of nature: the contrast between the regular and constant work of nature and the extraordinary variety of her forms results in a "living harmony." Consequently, art's main achievement is not to represent the forms of nature—i.e., what is observable—but to recreate and make observable the very mind of nature: "the mind of the painter must transmute itself into nature's own mind and became the interpreter between art and nature" (Leonardo, *Trattato*, I, 36).

The representation space of Renaissance painting is compared to the representation space of quantum physics in Chap. 6. The action of a "semi-transparent" mirror splitting the trajectory of a photon, or any quantum particle, resembles the action of "Alberti's window" in painting. According to Alberti, painting must recreate a view through a window. To accomplish this task, the light coming from the scene (to be painted) must be caught by the painter's eye and projected on the plane surface of an ideal veil. The pictorial image results from a "double projection," for that window acts both as a glass intersecting the visual pyramid and as a mirror reflecting the painter's eye. When Narcissus realizes how a mirror acts, the shadow which he observes allows him to see another side of himself. When physicists realize how quantum interference acts, "photon-shadows" become "observables." The awareness shared by Narcissus, Renaissance perspectivists, and quantum observers involves a revision of the function of shadow with respect to the Platonic condemnation and to the "fantastic" conception of the classic world. In the girl's attempt to fix her lover's image (told by Pliny), one can already see a "measuring her self against the other" that Plato's myth had not envisaged. Despite the separation from the young man, the image she drew, the "artificial" representation, would maintain the link (of her self) with the "other." This view of painting giving shape to "relational" forms marks itself off from a view of science describing "objective" properties of physical reality, hence losing or denying any link between the scientist-observer and the observed object.

A shadow—like that which Pliny's young woman wouldn't like to separate herself from or which Ovid's Narcissus would like to join himself to—is but an image: it is not a body, but a *projection* of a body. And it is on projections that painting works. Here lies its misery and its greatness: misery, for it can solely draw an image of a body's shadow; greatness, for it can recreate a "body" from shadow. The *double* projection involved in painting implies a double deception, or one might also say a double negation, i.e., an "affirmation" of deception. In this regard, Vasari's attempt to draw his self-portrait through the outlined shadow is eloquent (Fig. 1). It was a vain attempt, for the resemblance of a person's face is manifest in the form of a profile, and a self-profile cannot be directly observed, nor reproduced.[1] Nevertheless, this limit of "observability" unveils a peculiar character of the post-Albertian conception of art reviewed by Vasari. As the painting of the Cinquecento combines the action of shadow—the "relational" form of symmetry— and the action of mirror—the "static" form of symmetry—into an ideal harmony, so quantum physics combines "shadow-images" and "mirror-images" brought about by its "incompatible observables" into a complex mathematical architecture. Moreover, even Narcissus' disenchantment (Fig. 2) is brought about by a process of "measurement," i.e., by his vain attempt to "interact" with the *other*: the pleasure of sight does not end in the pleasure of the embrace. Indeed, *soli non possumus ludere*, as Cardano observes in his pioneering work on probability, *Liber de ludo aleae*.

While artificial perspective brings to mind a variety of artists, such as Brunelleschi, Alberti, Piero, and Leonardo, complex numbers and probability evoke the same "artist," Girolamo Cardano. It may well seem ironic that, already in the sixteenth century, one gambling scholar was concerned with concepts and methods which would become the primary architectural elements of quantum physics. At that time, however, artists and mathematicians did not use their knowledge in the context of official science, but rather in the epistemologically looser context of *art*, either the art of painting or the art of computing and gambling.

---

[1]Cf. V. Stoichita, *A Short History of the Shadow*, London: Reaktion Books Ldt. 1997, chap. 1.

Fig. 1 Giorgio Vasari: *The Origin of Art*, *c*. 1572. Florence, Casa Vasari. (Detail of the self-portrait.)

Though the "scientist" Cardano seemed reluctant to speculate over "sophistic roots" emerging from the solution of cubic equations, the "philosopher" was motivated to explore their *subtilitas*. As if he were concerned with Plato's dialog, Cardano uses the attribute "sophistic" in describing the "negative roots." These roots, later called "complex numbers," seemed to have such an ambiguous relation with "real" things (*aletheia*) as to appear a product of that art that the Eleatic Stranger in *The Sophist* defines *tékne phantastiké*. Cardano stresses their *inutilitas* when he describes them as *surdae*, meaning "non-natural," or *remotae a natura*. Consequently, a mathematician defending their utilitas within the natural order would lie, aware of lying. For Cardano, however, *inutilitas* does not hold a negative connotation, *tout court*; it is intimately related to that *subtilitas* which constitutes the essential character of his scientific-philosophical system and gives name to one of his main writings.

The origins of complex numbers in the late Renaissance are examined by Veronica Gavagna (Chap. 7). The *radices sophisticae* appear in the chapter of Cardano's *Ars magna* (1545) devoted to investigating the existence of "false" roots of a quadratic equation. However, Cardano did not attribute any sensible meaning to such strange mathematical objects. It was Bombelli who established their complete mathematical legitimacy, by means of a "geometrical" representation of the roots of an *irreducible* cubic equation. But Bombelli's insights were largely ignored by the European mathematical community of the late Renaissance: "complex numbers" were solutions to "algebraical" problems. Those sophistic quantities re-appeared on the scene in the form of Girard's *solutions impossibles* or Descartes' *racines imaginaires* to ensure the validity of the so-called Fundamental theorem of algebra.

Fig. 2 Leonardo da Vinci:
Narcissus, c. 1490. London,
National Gallery

As Renaissance mathematicians had to go through the uncharted territory of
complex numbers in order to obtain real solutions to cubic equations, so do quantum
physicists. Artur Ekert explains how complex amplitudes are used in order to
calculate probabilities. The connection between amplitudes and probability is not
trivial. His essay (Chap. 8) shows that any good statistical framework theory, a
meta-level description of the world, requires complex numbers. In particular, it is
argued that once we request continuity of admissible physical evolutions we will
end up with quantum theory, and if this requirement is dropped, we obtain classical
probability theory. Thus quantum theory can hardly be any different from what it is,
and Cardano's "useless" quantities can hardly be avoided.

Chapter 6
Artists and Gamblers on the Way to Quantum Physics

Annarita Angelini and Rossella Lupacchini

Quantum physics does not describe the world as the ultimate "artefact" of God acting as a creator and ruler. If God played a role in the quantum world it would be as a gambler. This quantum imagery was utterly unacceptable to Einstein, as reported in a famous letter to Max Born:

> In our scientific expectations we have progressed towards antipodes. You believe in the God who plays dice, and I in complete law and order in a world which objectively exists, and which I, in a wildly speculative way, am trying to capture.[1]

At first glance, the perspective attributed to Born, and to quantum physics in general, seems rather disturbing. Upon reflection, that perspective appears epistemologically fruitful and external to any scientific or theological dogmatism. Consider the game of dice. In order to make it fair (*aequus*), certainty must give way to probability. In order to make it possible, God ought to descend into the world, or the world to be elevated to the "divine". Indeed "*soli non possumus ludere*", as Cardano captured in his *Liber de ludo aleae*.

The doctrine that certainty is impossible, hence probabilities must be relied on, is named "probabilism" in the Renaissance:

> Probabilism is a token of the loss of certainty that characterizes the Renaissance, and of the readiness, indeed eagerness, of various powers to find a substitute for the older canons of knowledge. (Hacking 1975, p. 25.)

In line with those canons, nature is the "written word", the writ of the Author of Nature. Ian Hacking focuses on the fact that the Renaissance faced "probability" as an attribute of the *opinion*, whereas *knowledge* could only be obtained "by demon-

---

[1] Letter to Max Born, 7 September 1944.

A. Angelini (✉) • R. Lupacchini
University of Bologna, Bologna, Italy
e-mail: annarita.angelini@unibo.it; rossella.lupacchini@unibo.it

R. Lupacchini and A. Angelini (eds.), *The Art of Science*,
DOI 10.1007/978-3-319-02111-9_6,
© Springer International Publishing Switzerland 2014

stration". This probability was not "support from evidence" but from authority: *signs* had probability because they came from the ultimate authority. The Renaissance "scientists" were after knowledge and demonstrative science, but demonstrations were out of reach for physicians, alchemists, astrologers, and magicians. *Scientia* described the "real" world by universal truths, but the Renaissance physician had to prescribe and predict from "probable" signs collected through phenomena. These signs, therefore, had probability because they came from nature, not from the writ of its Author.[2] One of the most skilful physicians of the Renaissance, Girolamo Cardano, was also a gambler, and his *Liber de ludo aleæ* is credited as the first book on probability.

Four centuries later, probability became magically involved with *complex numbers* through the notion of "complex probability amplitude" introduced by quantum theory.[3] Ironically, both probability and complex numbers draw attention to the work of Girolamo Cardano. Indeed, like probability, complex numbers cropped up in the Renaissance, and Cardano's *Ars Magna* is conventionally hailed as their birth certificate. Both subjects, however, were treated with a great deal of suspicion by the scientific establishment and, as a result, were overlooked for many years. Quantum theory establishes an original and peculiar link between them, but the meaning of that link remains rather mysterious.

Aware of the fascination pervading every mystery, this essay is not after solutions. Instead, it will focus on certain intellectual requirements in the culture of the Renaissance which would take a voice much later. As late as 1831, in his book *On the Study and Difficulties of Mathematics*, the logician Augustus De Morgan commented: "We have shown the symbol [square root of a negative number] to be void of meaning, or rather self-contradictory and absurd [...] Nevertheless, by means of such symbols, a part of algebra is established which is of great utility". De Morgan's remark echoes Cardano's, three centuries earlier, branding a square root of a negative: "*adeo est subtile, ut sit inutile*" (Cardano 1545; 2011, p. 243). Yet those useless roots push complex numbers into existence, and complex numbers enable quantum theory to take shape. But what part is played by complex numbers and probabilities in scientific knowledge? Did they appear in the Renaissance as unexpected gifts of chance, or as eloquent expressions of a refined vision of *scientia*? A vision which Renaissance art was ready to grasp and contemporary science still tries to accommodate.

6.1   The Art of Play and the Science of Art

Our historical and intellectual starting point is the humanism of the *quadrivium*, a "scientific" humanism characterized by patterns of cultural renewal and discontinuity with respect to the logical and epistemological late-scholastic tradition.

---

[2]According to Hacking, "What happened to signs, in becoming evidence, is largely responsible for our concept of probability" (1975, p. 35).

[3]See Ekert's essay in this volume.

The involvement of arts, in particular the arts of drawing (painting, sculpture, architecture), with scientific thought is the core of a *visio intellectualis* of the Renaissance as an age of "crisis". The crisis, which affects taxonomies and foundations of knowledge, conflates a number of different regions of learning. As a result, a variety of disciplines, old and new, from politics to painting, provides a laboratory for devising and testing new conceptual and scientific models.

A peculiar character of the Renaissance culture emerged from a symbolic conception of reality in the frame of a productive "imagination" of neo-Platonic ascendance. Experimenting entails questioning, directly and primarily, the conditions and the procedures according to which scientific knowledge is attained and corroborated. From this point of view, an enquiry into structural and conceptual relationships between the painters' science and the mathematicians' art becomes significant. Science and art are no longer confined to their respective cultural domains, nor conceived as impermeable to one another.

Without losing contact with objective reality, science and art try to withdraw from the plane of perception as much as from the plane of *noesis*, and stand apart. At the intersection between limitation and freedom, they tend towards a knowledge neither sensible nor noetic: a *dianoetic* knowledge, specular to the position of man in between.

Neither Aristotle of medieval metaphysics, nor Plato interpreted by Ficino, nor Cicero, nor Epicurus, nor Euclid could pave the way for a *dianoetic* foundation (we should rather say, a *dianoetic* suspension) of knowledge. Those traditions were branded by the assumption of the specularity between the scale of being and the scale of sciences, between the necessitarian character of a *scientia*, successful regardless of human constructions, and of a praxis oriented towards an end, regardless of the criteria of scientific proof. Here scientific knowledge neither denies, nor refrains from, a metaphysical dimension. Here art, more precisely the theory of art, appears mature enough to beget a paradigm beyond the domains and products of the arts of drawing.

The status of art (*téchne*), meaning freedom of not observing the principles of scientific logic,[4] allowed the construction of an imaginative reality, no evidence of which was to be found in either the immediate perception or ideal models, and yet was a *costruzione legittima*. This reality found its *auctoritas* within itself, and nowhere else. The point of view—consider, out of the metaphor, the one ruling linear perspective—was neither God's eye nor the mind's eye, but a reference system consisting of axes, points, and angles that enabled the artist to render the formal coherence and the perceptive effect of a reality first imagined, then depicted, and finally known.

The architect Filippo Brunelleschi is conventionally praised for the "invention" of *linear perspective*. His project of the cupola of Santa Maria del Fiore in Florence constituted a real challenge. What was striking for the Renaissance intellectual reflection was not so much the huge size of the cupola and the technical novelty involved in its construction, as the master plan of the architect. Indeed Brunelleschi was able to design a coherent and self-supporting structure, discharging weights and forces inside rather than outside, and to provide a demonstration of its holding by completing the job. With no prime principle, no a priori guarantee, no external fixed point—such as a centering (*centina*)—on which to lean, the effectiveness of the construction was claimed only at the end of a tentative, daring way.[5] This way had its incipit in the artist's productive imagination; his constructive hypotheses were discarded or taken step by step, and tested *post festum*—in the *prova della fabrica*—by the holding of the complete construction. Although Brunelleschi's masterpiece was hailed as an extraordinary new *inventum*, it was not meant to provide a unique experience. Since it could be encompassed by a rule (*ratio*), it posited a precedent for further experiments. A theoretical "map", with no evidence in concrete experience, guided the artist—like a gambler—to make conjectures and to calculate the probability for a *possible* outcome and the most effective strategy to accomplish it.

In the Chap. 14 of his *Liber de ludo aleae*, the gambling scholar Girolamo Cardano stated the following general rule:

---

[4]"The artists were self-taught and learned through practice. Fragments of Greek knowledge filtered down to them, but on the whole they sensed rather than grasped the Greek ideas and intellectual outlook. To an extent this was an advantage because, lacking formal schooling, they were free of indoctrination. Also, they enjoyed freedom of expression because their work was deemed 'harmless'" (Kline 1972, p. 231).

[5]As in Alberti's dedicatory letter of *De pictura* to Brunelleschi: "What man, however hard of heart or jealous, would not praise Filippo the architect when he sees here such an enormous construction towering above the skies, vast enough to cover the entire Tuscan population with its shadow, and done without the aid of beams or elaborate wooden supports?" (1436; 2004, p. 35). ["Chi mai sì duro o sì invido non lodasse Pippo architetto vedendo qui struttura sì grande, erta sopra e' cieli, ampla da coprire con sua ombra tutti e' popoli toscani, fatta senza alcuno aiuto di travamenti o di copia di legname, quale artificio certo, se io ben iudico, come a questi tempi era incredibile potersi, così forse appresso gli antichi fu non saputo nè conosciuto?" (1436; 1975, pp. 7–8).]

we should consider the whole circuit, and the number of those casts which represents in how many ways the favorable result can occur, and compare that number to the remainder of the circuit, and according to that proportion should the mutual wagers be laid so that one may contend on equal terms.[6]

The rule prescribing the possible outcomes of casting two dice provides a proper definition of the notion of probability as a ratio between favourable and possible cases. A theoretical condition, no evidence of which came from concrete experience, guided Cardano's conjectures about the game of dice: the ideal simultaneousness of all possible outcomes. Facing a purely theoretical situation, Cardano was able to calculate the probability for a particular outcome.

Alberti (1485; 1966, pp. 9–11) explained that art should not be judged on what it *is*, but on how it *works*. We accept it if effective (*fas*), and refuse it if it does not work, or does not work adequately (*nefas*). Cardano used probability calculus with a similar attitude. If a conjecture can be taken as a rule, it is not because the conjecture is a necessary law at the foundation of objective reality, but because the "things" of nature (*res*) come close to the conjecture of the subject. Although the outcome of a roll of dice depends both on the physical features of the dice and on the gambler's skill, it belongs to a world utterly different from both the world of the dice and the world of the gambler. That world is located in an intermediate space endowed with rotational symmetry, the ideal space of the *circuitus*, which plays host to all (simultaneously) possible outcomes.

Cardano's fundamental assumption was a notion of *aequalitas* which he cared to distinguish from the Aristotelian *mediocritas*. The latter comes from mediation and equilibrium between two extremes fixed, known, and certain. But in the game of fortune, the mean point is more certain than any extremes. If there is an equilibrium between fate and fortune, then it is fair to bet on the result of a roll of the dice. For Cardano, that equilibrium was nothing but chance.

Author of a dialectic attempting, as many others in the sixteenth century, to define one criterion of validity for the *humana sapientia*, Cardano, as a player, elaborated a "mathematical" theory of chance to entrust not to a "scientific treatise", but to a booklet with no scientific pretence. In its search for an answer to a given problem, the theory was designed to pursue a multiplicity of possible paths. By contrast, in making his commitment to a scientific treatise *stricto sensu*, namely the *Ars Magna*, Cardano seemed reluctant to pursue unlikely paths over "sophistic roots". Free to hazard unwonted strategies and new mathematical rules in the "art of play", he proceeded warily within the domain of an "art" whose scientific status and demonstrative power he did not intend to diminish or question: "this art surpasses all human subtlety (*subtilitas*) and the perspicuity of mortal talent and is a truly celestial gift and a very clear test of the capacity of men's minds" (Cardano 1545; 2011, p. 243).

---

[6]"Una est ergo ratio generalis, ut consideremus totum circuitum, et ictus illos, quot modis contingere possunt, eorumque numerum, et ad residuum circuitus, eum numerum comparentur, et iuxta proportionem erit commutatio pignorum, ut aequali conditione certent" (Cardano 1663a; 2006, p. 63).

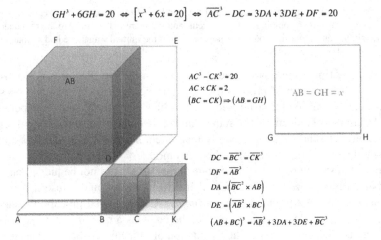

$$GH^3 + 6GH = 20 \iff \left[x^3 + 6x = 20\right] \iff \overline{AC}^3 - DC = 3DA + 3DE + DF = 20$$

Fig. 6.1 Cardano's geometrical proof of the case $x^3 + 3x = 20$

6.2 The "Great Art"

Cardano's *Ars Magna* was famed for exhibiting publicly, for the first time, the rule which solves "the cube equal to first power and a constant".[7] The rule, credited to Scipione Del Ferro of Bologna as well as to Niccolò Tartaglia of Brescia, was handed to Cardano by Tartaglia, though without demonstration.[8] Nevertheless, Cardano understood that the rule had been discovered through a geometrical demonstration and succeeded in setting down a version of his own in three dimensions. As it is shown in Fig. 6.1, the demonstration is drawn from the solid figures corresponding to the constituent parts of the cube of a binomial.[9] For Cardano, "this [the geometrical one] would be the royal road to pursue in all cases" (Cardano 1545; 2011, p. 107). But in fact, this does not work for the so-called irreducible case which calls "complex numbers" into play. Mentioned in passing by the *Ars Magna* (Cardano 1545; 2011, pp. 147–152), the case, which arises when the cube of one-third the coefficient of the first power is greater than the square of one-half the constant of the equation,[10] appeared irreducible to that of "geometrical terms".

---

[7] A cubic equation such as $x^3 = 3px + 2q$ is solved by the rule:

$$x = \sqrt[3]{q + \sqrt{q^2 - p^3}} + \sqrt[3]{q - \sqrt{q^2 - p^3}}. \tag{6.1}$$

[8] For a more detailed presentation, see Gavagna's essay.

[9] Such as $(x + a)^3 = x^3 + 3x^2a + 3xa^2 + a^3$.

[10] Namely when: $p > q^2$.

Indeed, for Cardano as well as for most mathematicians of the Renaissance, the role of "algebra" was still ancillary to geometry:

> Although a long series of rules might be added and a long discourse given about them, we conclude our detailed consideration with the cubic, other being merely mentioned, even if generally, in passing. For as *positio* [the first power] refers to a line, *quadratum* [the square] to a surface, and *cubum* [the cube] to a solid body, it would be very foolish for us to go beyond this point. Nature does not permit it. (Cardano 1545; 2011, p. 72)

Consequently, it seemed foolish for them to venture into "mental tortures" (*incruciationibus*) such as negative square roots. Despite various attempts to attribute a meaning to the "sophistic roots" involved in the *casus irriducibilis*, neither Cardano nor Tartaglia was able to come to terms with them.

According to Greek-Euclidean tradition, the geometrical forms establish trustworthiness to mathematics; in the Renaissance, the very rules of "algebra" seemed willing to secure a higher level of generality to geometrical forms. These rules make it clear that the meaning of the numerical "symbols" and operations cannot simply be drawn from immediate intuition, but requires a space free of presuppositions to unfold. Bombelli's *Algebra* can be regarded as such a space, welcoming "sophistic roots" as a new kind of numbers to be equipped with the appropriate rules.

> Another instance of "conjugate cubic roots" (*radici cubiche legate*) originates in the chapter on the cube equal to the first power plus a constant, when the cube of one-third the coefficient of the first power is greater than the square of one-half the constant [...]. This kind of square root has, in its algorithm, an operation different from any other and a different name; for, when the cube of one-third the coefficient of the first power is greater than the square of one-half the constant, what exceeds cannot be called plus nor minus, it will be named "plus of minus" if it is to add, whereas it will be named "minus of minus" if it is to subtract. And this operation is "very necessary"... as so many are the cases bringing about this kind of root [...] which will appear rather more sophistic than real, and such was also my own opinion until I found its demonstration in lines (as it is shown in the demonstration of this chapter on a plane surface).[11]

Jacob Klein (1968) questions which transformations mathematics had to undergo in order to let a symbolic modern algebra grow out of the "geometric" algebra of the Greek scholastic tradition. The mathematical disciplines traditionally belonged among the *artes liberales*, intended as theoretical disciplines in contrast to the practical *artes mechanicae*. Arithmetic, however, maintained close links with the "art of calculation", for its "logistic" elements provided the theoretical foundations

---

[11]"Un'altra sorte di R.c. L molto diverse dalle altre nasce dal capitolo di cubo uguale a tanti e numero quando il cubo di un terzo delli tanti è maggiore del quadrato della metà del numero, come in esso Capitolo si dimostrerà, la qual sorte di radici quadrate ha nel suo Algoritmo diversa operazione dalle altre e diverso nome; perché quando il cubato del terzo delli tanti è maggiore della metà del numero, lo eccesso loro non si può chiamare né più né meno, però lo chiamarò più di meno quando egli si doverà aggiongere, e quando si doverà cavare lo chiamerò men di meno, e questa operatione è necessarissima [...] che molto più sono li casi dell'agguagliare dove ne nasce questa sorte di radici [...] la quale parerà a molti più tosto sofistica che reale, e tale opinione ho tenuto anch'io, sin che ho trovato la sua dimostrazione in linee (come si dimostrerà nella dimostrazione del detto Capitolo in superficie piana)" (Bombelli 1572; 1929, pp. 133–134).

for "practical" calculations. According to Klein, in the Renaissance, the "artful" character of *all* mathematics rooted in the original kinship of *téchne* and *episteme* was slowly identified with "practical application", in the sense of the application of a skilful *method*.

> These disciplines are consistently understood as "*artes*"; to learn them means to master the corresponding "rules of the art". [...] Now at that significant moment when these disciplines first succeed in gaining recognition as part of the "official" science, it is precisely their character as "arts" which is thought to lend them their true theoretical dignity. (Klein 1968, p. 125)

At that moment, a *new* kind of "symbol-generating abstraction" paved the way for modern algebra.

The novelty advanced along two main routes: one can be traced to Greek sources, in particular to Diophantus' *Arithmetic*; the other, from the Arabs, carries with it independent pre-Greek elements beside the Greek sources (Klein 1968, pp. 147–149). A symbolic technique of counting seems to flow very naturally from the *Arithmetic* of Diophantus. Here, operations involving numbers of different kinds, except for negative, are carried out with ease, and the concept of *eidos* is used in a purely instrumental way. Though all this reveals a doubtless inner tension between the matter treated and the character of the concepts forced on it, it did not push its way any further. Concerning the algebra drawn from Arabic sources, despite its techniques of calculation were continually elaborated, so far as to introduce "negative", "irrational", and even the so-called imaginary magnitudes (numbers *absurdi* or *ficti*, *irrationales* or *surdi*, *impossibiles* or *sophistici*), its self-understanding failed to keep pace with these technical advances.

By emancipating algebra from spatial intuition, a cubic equation can be viewed in two dimensions. Bombelli's demonstration "on a plane surface" provided a general statement of the existence of real roots of a cubic equation, even when the irreducible case occurs. In this case, the cube cannot be resolved into its "bodily" components, yet can still be deconstructed through an "imaginary" projection on a plane. The case $x^3 = 6x + 4$ is sketched in Fig. 6.2. Separated from the traditional scholarly disciplines, the algebra which proceeds from Fibonacci (Leonardo of Pisa) and the abacus masters of the thirteenth century *via* the school of the *cossisti*, struggled for a place in the system of "science". This algebraic school—which flourished within the realm of "low sciences"—became conscious of its own "scientific" character and of the novelty of its "number" concept only at the moment of direct contact with the corresponding Greek science, namely the *Arithmetic* of Diophantus. Bombelli was the first mathematician of the Renaissance who assimilated Diophantus' work.[12] It was in contrast to the *geometric* algebra

---

[12]After studying and translating the first five books of the Diophantine manuscript, Bombelli changed the form of his 1550 manuscript (Bortolotti 1929).

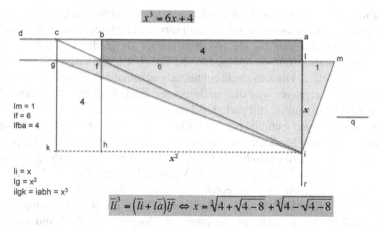

Fig. 6.2 Bombelli's proof on a plane surface. The parallelogram "ilg" is the cube x^3, while the parallelogram "ilf" is $6x$, for "il" is x and "lf" is 6; the parallelogram "hfg" is 4, because it is equal to the parallelogram "alf", which is 4 because "ilg" equals $6x$ plus 4. Hence x^3 equals $6x$ plus 4

of Diophantus that the new *linear* character of Bombelli's algebra became visible. Nevertheless, the significance of Bombelli's work remained invisible to most mathematicians.

6.3 In the Light of *Subtilitas*

The age of Brunelleschi, Cardano, Bombelli, and the *perspectivi* marks the beginning of a passage from an ontological conception of the world to its symbolic representation. This passage, which appears today at the origin of a science of possibilities and relations, in contrast to a science of beings and substances, brought the "problem of form" to the fore (Cassirer 1927, p. 143). It was a problem faced by artists, in particular painters, with a new sensibility. In performing a creative activity, the artist dismissed any "copiative contemplation of the datum" and recognized, as a natural law ruling their works, nothing but the freedom of seeing and of giving form: the "natural necessity". When natural philosophy was still unable to get rid of a notion of natural law grounded on ontology, and when algebra was still subordinate to a geometry anchored to a physical and metaphysical space, the theory of art defined anew the problem of form overtaking the subtleties of late Scholastic logic.

Art preceded philosophy in the discovery of a different scientific legitimation and a new conception of nature. The point at issue is not whether art, a free creative activity, was able to become a "science of art" following objective, necessary, and compulsory mathematical laws, but rather how theory of art and theory of science, tied by a new cognitive relationship, were able to shake the logical and ontological structures of the past tradition. In the fifteenth and sixteenth centuries, artists were the first to reassess the terms of the relationship between subject and object, self and nature, liberty and necessity. Painters, sculptors, and architects, as well as natural magicians, gradually saw that the object of human knowledge and activity is not something separated from-or opposed to-the self, but is the target of all creative energies of the self. A virtuous circle connected mathematical and artistic inventions in a common descriptive, symbolic, architectonic goal. And these inventions drew from art their legitimation, still partial and transitory.

Some artists played a crucial part in this transition, that can hardly be reduced to the contrast between Medieval fantasy and modern freedom (Kline 1953, p. 100). This is the case of Brunelleschi, mentioned above, for his astonishing enterprise in the construction of the cupola of S. Maria del Fiore. This is also the case of Bombelli, whose *Algebra* accepted the same vertigo of sense introduced in architecture by Brunelleschi a century earlier. With an attitude similar to the one adopted by the architect of the dome of Florence, Bombelli did not elude the void, the lack of sense which threatened the "conjugate cubic roots". As an artist, he faced that "wild void" and found his way through it: performing an operation "different from any other", accepting that diversity. Artists such as Brunelleschi and Bombelli, as well as Piero della Francesca and Leonardo, may appear so "modern" in their respective achievements as to suggest the idea (almost always wrong) of a transition from the old to the new in terms of a choice between alternative options. Others, like Cardano, better illustrate the complexity of the transition, which is of interest not only when it provides successful results, but even more so when it shows contradictions, uncertainties, and misfortunes.

"Cardano," writes Morris Kline (1953, p. 122), "bridged the gap between the Middle Ages and modern times". Philosopher, mathematician, astrologer, magician, gambler, and professor emeritus of medicine at the University of Bologna, Cardano embodies the model of the sixteenth-century intellectual struggling between the official, traditional knowledge and the emerging (or re-emerging) civil, humanistic culture. He was aware of playing both roles as a protagonist. Cardano was an artist, a practician like Brunelleschi, when writing about commercial mathematics or gambling, he was a "civil" humanist when dealing with morals, history, dialectic, and, finally, he was a *philosophus* when writing about geometry, medicine, and the meta-physical constitution of the universe and human nature. Still, he mixed up the cards of classical taxonomies of knowledge to such an extent as to use Latin both in *De sapientia* and *De Ludo Aleae*. Unlike Bombelli, Cardano "is not bold enough to accept imaginary numbers" (Koyré 1958, p. 32). Yet, not resigned to the "irreducibility" of one case, he endeavoured to demonstrate the general rule of cubic equations. He needed a reliable procedure to follow, since he well understood, like Brunelleschi, and better than Tartaglia, that *scientia* is nothing but knowing how to operate according to a rule.

Aiming at raising unexpected results to the level of valid knowledge, Cardano's *scientia* was not the discovery of an absolute first principle, but the construction of a transmissible and repeatable procedure. Evidence of this is provided by his theory of chance, as well as by his dialectic, medical, and astrological conceptions.

His obstinate search for a rule and his acceptance of a criterion of approximation (provided that approximation be "measurable") illuminate the most proper meaning of that reform, or reformulation, of the notion of scientific legitimacy which marked the crisis and the theoretic contribution of the Renaissance. Through the practice, i.e. the art, of gambling, Cardano recognized that knowledge is not fallacious (*non fallit*), even when it rests on conjectures and approximation (*secundum coniectura et proximoirem*); he experienced that, although a conjectural ratio could not be assessed as a truth, usually "things happen getting close to the conjecture" (*res succedit proxima coniecturae*) (Cardano 1663a; 2006, p. 57). For the established sixteenth-century philosophy and science, a scientific law conveyed truth deducted *ex causas*; a law revealed its necessity by establishing, through reasoning (*syllogismus*), a truth grounded on certain premises metaphysically guaranteed. However, the rule used by Cardano to calculate the possible outcomes of gambling and to give scientific legitimacy to singular cases has no metaphysical foundation. It is not the scientific truth that adheres to the metaphysical (universal and necessary) truth, but it is the singularity of cases that approximates—in the range of possibilities—to the generality of a rule of reason. Therefore, to know is not to contemplate the truth, but to master rationally the distance between the object to be observed and the eye of the human observer, as did painters to give the appearance of a third dimension on a flat surface, transforming—on purpose—the rules of geometric optics.

Nevertheless, Cardano was not able to find a way to legitimate the sophistic roots involved in Del Ferro and Tartaglia's formula. Those roots were "sophistic" because they were inconsistent with the nature of line and surface, an ingenious and semi-divine discovery, but also as *subtilis* as *inutilis*. Facing that kind of roots, Cardano withdrew, as he could not cope with them. He could not see the negative side of a square which, of course, does not exist. In *Ars Magna*, his most important mathematical work, he was not capable of doing what he preached in *De sapientia* and *De Subtilitate*: mastering—simulating, managing, approximating, "making to appear"—that third dimension which separates the eye of the geometer from an entity which eludes the geometric reality. For Cardano, imaginary numbers were *surdi*, meaning not that they are absurd, or *alogoi*, but that they are rather non-natural, or *remoti a natura*. Even the adjective "sophistic" should not be interpreted in solely pejorative terms. Enemy of the truth is not the sophist, but is Socrates, who dissimulates ignorance and deception by preaching an alleged absolute truth. Masters of rhetoric assess deception and approximation as inherent in "human knowledge"; real impostors and false *predicants*, in Cardano's opinion, are those who present as natural what instead derives from human passions, desires and limits, and those who present as indubitable and divine solutions that are not less false than those of the sophists and artists, but are kept exclusive and used as a means of subjection and power instead of being made available to all.

It seems reasonable that the mathematician maintained what the philosopher claimed in *De sapientia*: chased out of Eden after the Sin, man was separated from truth (Cardano 1544; p. 493), and since then, human knowledge tends towards *rationes verae*, stored in a secluded paradise, which can be approached, but never reached. Because of his ability to dissimulate a distance which cannot be bridged, man is the only "animal able to deceive", that is to say, able to present or represent, like an artist, what appears as what is. The danger is not in the limit, or in the appearance of truth, nor in the conjectural character of "human knowledge", but in presenting as true what is not true. Cardano was not interested in elevating what is false to what is true, but in showing that behind a "sophistic human knowledge" there is a "reality" not necessarily true, and yet not false. In the light of the lexicon of *De sapientia* or of *Encomium Neronis*, the sophistic roots are signs of an undoubtably human and limited knowledge, but not—for this reason—false or not legitimate. They mark a limit which, according to Cardano, is movable and never final, just like the limit of the *artes*, whose aim is to cover the distance from that pure and absolute truth which remains unaccessible to Adam's progeny (Cardano 1550, p. 551).

Surdae and *sophistic*, and also so *subtiles* to be *inutiles*: *"adeo est subtile, ut sit inutile"* (Cardano 1545; 2011, p. 242). It is even more difficult to give a negative connotation to a *subtilitas*-a name which Cardano gave to one of his most important works (*De subtilitate* 1550)-meaning something ambiguous, objective, and subjective at the same time. *Subtilitas* is the elusive web of a universe interwoven with imperceptible, unstable, moveable elements, continuously transforming and reciprocally communicating. *Subtilitas* is the character of an intelligence able to seize the remotest and deepest relationships, to discern the smallest parts, and therefore to grasp the *subtilitas* of the nature itself. Thus, it is the *subtilitas* that ties nature (object) with the thought (subject) penetrating it, that allows human freedom to interfere with natural necessity. Through the notion of *subtilitas*, Cardano focused on the minimality of causes and the finiteness of principles, to conclude that reality is not what the eye sees, but what intelligence decomposes and discriminates. Those subtle roots then stand on the edge, separating and connecting the imperceptible web (remote from sense) of the universe to the powerful lens of a *subtilis* intelligence, able to see the smallest parts of that web.

Negative roots are *subtilis*, and yet *inutilis*, in *Ars Magna*, but they are not mentioned in the Book 15 of *De subtilitate* concerning "useless inventions". They are *inutilis* because they have no use: they cannot be applied to a mathematics which connects the notion of number with a geometric entity. *Surda* and *sophistic*, *subtilis* and *inutilis* are pairs of adjectives that bring to light a more general issue, i.e. the discontinuity of reality, which convinced Cardano the philosopher, but not Cardano the mathematician. The procedures transforming the conic sections, described in the Book 16 of the *De Subtilitate* (later borrowed by Desargues), confirm the hypothesis of a continuous progression. Nevertheless, the physical and meta-physical universe of *De Subtilitate* disproves that continuity, showing instead fractures, however small, in the web of the universe: void points and deep shadows, unable to comply with the geometric continuum and, consequently, of no use (*inutiles*) in mathematics. The point is not that Cardano was not bold enough to see the meaning of imaginary numbers, but that he was unable to disentangle algebra from geometry. The point concerns Cardano's mathematics—as, more generally, sixteenth-century mathematics—whose *subtilitas* was not so refined as the *subtilitas* of his "natural" universe. *Felix culpa*, since Cardano seems to recognize it and to point a possible way out.

6.4   The Science of Shadows

In *De Subtilitate*, it is painting that compensates for the faults of geometric optics, being better equipped to grasp and represent that shadow, void, or absence which geometry eludes. The geometry of vision fails to represent phenomena perceived in three dimensions; its laws describe only the front surface of a solid. What allows us to distinguish a square from a cube is not a flat vision to be rendered by means of angles and lines of plane geometry, but a judgement (*iudicium*) drawn from past experience and adapted to the geometrical form of the visual perception (*iudicio ex diuturno usu contracto*) (Cardano 1550, p. 428). Our judgement is also triggered by shadow, which is as crucial as light in "demonstrating" a third dimension which does not impress the eye and cannot be forced on a plane. Of the things before us, we *judge* as "real" both what we actually see and what we do not see, but conjecture hidden somewhere in the shadow.

For Cardano four elements were needed to represent three-dimensionality: the shape of the object (inferred from visual lines), the shadow, the colour, and the relationship between the observer's point of view and the object of the representation; only the first can be expressed in geometrical terms. In contrast with this *diminutio* of geometry, which is unable to benefit from colour and shadow, painting adds a dimension, even with respect to visual perception. By tying things to a subject's judgement, painting is able to see and show, although by conjectures (*coniecturari*), what our eyes cannot see. In his view of painting, Cardano seems to move away from the laws of *Ars Magna* and *Encomium geometriae* and proceed towards the conjectures and rules of *De Ludo Aleae*. Here, Cardano's conception of knowledge appears both demonstrative and conjectural.

Painting does not reproduce the object, but imitates "the affections of the object". For instance, in providing a representation of a human body, painting represents not only shapes, colours and proportions of the body, but also "conjectures" about dispositions of the mind. These invisible matters are made visible through the affections of the body (Cardano 1550, p. 428). For making visible what is lighted as well as what is shadowed, and hence for endowing a judgement with a *form*, painting is esteemed as the *subtilis* art *par excellence* (*subtilissima omnium*) (*ibid.*, p. 609). By adding the colours and the shadows of its fabric to the rules of geometrical optics, painting is able to bring about and legitimate new inventions. As Leonardo wrote:

> Such a proportion is between the imagination and the effect as between the shadow and the shadowed body, and the same proportion is between poetry and painting, since poetry set

Fig. 6.3 Leonardo's illustrations for Pacioli's *De Divina Proportione* (1509)

its things in the imagination of letters, and painting set them out of the eye from which it receives the similitudes as if they were natural.[13]

To a geometric line repelling a "negative" nature, painting seems to add a "*lineamentis*" to be traced back not so much to Euclid as to Alberti's *lineamentum* or Zuccaro's *disegno interno*.[14] Cardano's view of painting conjugates the objective *subtilitas* of the natural world with the subjective *subtilitas* of an intellect able to discern it. If a geometer is a theorist (as a mathematician), then a painter is not only a practician (as an artist), but also a philosopher, an architect, and an anatomist. Cardano saw this exceptional figure of "sapient" embodied by Leonardo: both theorist and practician, mathematician and artist, philosopher and architect.

Indeed the *subtilitas* of solid forms is masterfully rendered by Leonardo's illustrations of the *De Divina Proportione* by Luca Pacioli (1509). To let painting unfold into a mathematical vision, Leonardo depicted a series of polyhedra, which Pacioli named "*dependenti*", besides the five Platonic solids (tetrahedron, cube, octahedron, dodecahedron, icosahedron). With respect to coeval illustrations of Platonic solids, Leonardo's polyhedra (Fig. 6.3) appear in a perspectival configuration,

---

[13]"Tal proporzione è dalla immaginazione all'effetto, qual è dall'ombra al corpo ombroso, e la medesima proporzione è dalla poesia alla pittura, perché la poesia pone le sue cose nella immaginazione di lettere, e la pittura le dà realmente fuori dell'occhio, dal quale occhio riceve le similitudini non altrimenti che s'elle fossero naturali" (*Trattato*, I.2).

[14]For more details, see Vesely's essay.

Fig. 6.4 Attributed to Jacopo
Caraglio, *Diogene, c.* 1525.
Florence, Gabinetto disegni e
stampe degli Uffizi

which emphasizes their spatiality and visualizes them not as abstract entities, but
rather as objects of experience.

The plastic impression, rendered through the shadow and the tension of the
ribbon holding an object endowed with weight (not a pure geometric object), reflects
the urge to confer the same "objectivity" of the Platonic solids to innumerable
dependenti solids,[15] so that the latter, just like the former, can be treated as if they
were "natural". As Cardano stated in his *De Subtilitate*, the painter must observe,
compare, and measure the magnitude, quantity, form, colour, motion, cavities, and
all other innumerable details. Thus, he must conceive, in his mind and even in his
memory, what he has seen before; next, he must delineate the *typus subtilis* of each
part separately, first depicting each of them as singular, then all together as a set, so
as to render the symmetry between parts (Fig. 6.4).[16]

The "virtuous circle" between artists and mathematicians become more and
more effective. Thanks to Leonardo, the regular polyhedra described by Pacioli
became *"subietto della virtù visiva"* and, therefore, "knowledge able to reach all

---

[15]Although the polyhedra drawn by Leonardo are in twenty-eight tables, *De Divina Proportione*
does not put an upper limit on their number.

[16]"Primum quidem, quia generaliter duplus est labor, inde comparatione, si quid artifex delituit
in magnitudine, numero, forma, colore, lituris, rugis, cavitatibus, aliisque innumeris, quae in unius
medietatis figura lacebant, manifesta facta, operis turpitudinem declarant. Qui igitur fingere aliquid
volunt, formam eius primum visam mente, quasi memoria concipere debent, inde typum quendam
seorsum delineare subtiliu, post praesente eo quod signis singula animadvertendo ad amussim
perficere, latet enim in unoquoque partium quaedam symmetria, quam si non mente conceperis
oculorum vero praesidio tantum innixus tentes exprimere, operam luseris" (Cardano 1550, pp. 609–
610).

generations of the universe". But did Leonardo's painting serve only to diffuse Pacioli's mathematical knowledge?

Pacioli's knowledge about regular polyhedra was drawn from the *Trattato d'abaco* and the *Libellus de quinque corporibus regolaribus* by Piero della Francesca.[17] Piero tackled the geometrical problem in the frame of a series of questions of stereometry to be solved by means of arithmetic and algebra. Any mystical or philosophical meaning is set aside from his analysis of polyhedra. These solid bodies were not copies of models of an ideal geometry, but they possessed self-referentiality with respect to the sensible and metaphysical order. Accordingly, he trusted not geometrical lines and angles to work them out, but instead numbers and roots. It is from his painting experience that he drew this kind of "symbolic" formalization, as it cannot be composed of simply sensory data or the geometrization of those data. Piero addressed the problem of form in the manner of an artist both in his painting and his mathematics. This leads back to his treatise on perspective (Piero della Francesca 1474). Seeing geometry as an art, Piero derived innumerable regular polyhedra from the five Platonic solids, and treated the five Platonic solids—"as many and sufficient as required by Nature"—as a subset of the innumerable *generable* polyhedra. Finally, Leonardo used the "science" of painting to give life to both. As we read in his *Trattato della Pittura*,

> [The eye] triumphs over nature, in that the constituent parts of nature are finite, but the works which the eye commands of the hands are infinite, as is demonstrated by the painter in his rendering of numberless forms.[18]

The regular polyhedra studied by Renaissance mathematicians and drawn by Leonardo are invariant under certain rotations in three-dimensional Euclidean space, or, in more sophisticated mathematical terms, are invariant under a "subgroup of the special orthogonal group" (Weyl 1952, p. 99). For Leonardo then, the problem at hand was how three-dimensional rotations could be "projected" onto two dimensions. After four centuries, an analoguous problem is posed to quantum theory by its "incompatible observables" and solved in a *complex* space. Of course, Renaissance mathematicians could not even have dreamed of a complex space where the rotation symmetries of their regular polyhedra could acquire a physical meaning. Yet, those symmetries are exceptionally well rendered by Leonardo's tables. Indeed, just when mathematics was reluctant to accept "imaginary" numbers, by opening the figurative space to an imaginary dimension, artistic *vision* "championed the rights of scientific abstraction and paved the way for it" (Cassirer 1927, p. 158).

---

[17]For a detailed account of Piero's art and mathematics, see Field (1997; 2005).

[18]"Le opere che l'occhio comanda alle mani sono infinite, come dimostra il pittore nelle finzioni d'infinite forme" (*Trattato*, I.24).

6.5   The Science of Painting and the Art of Measuring

According to the mathematical idealism of Leonardo and Galileo, mathematics provides criteria for a systematic construction of nature. Mathematics explains how to separate the "necessary" (that which obeys laws), from the "accidental" (that which is fantastic and arbitrary). But "the issue was not decided by purely intellectual [motives] alone". As Cassirer emphasizes, "[in] a manner which is characteristic and determinative of the total intellectual picture of the Renaissance, the logic of mathematics goes hand in hand with the *theory of art*. Only out of this union, out of this alliance, does the new concept of 'necessity' of nature emerge" (Cassirer 1927, p. 152). Mathematics and art now agree upon the same fundamental requirement of "form". And yet, while mathematical certainty has a unique form, painting presents us with infinite forms.

Although both Leonardo and Galilei saw in the "necessity" the decisive character that distinguishes what we call "nature" from what is a product of imagination (*fantasia*), according to Leonardo, unlike Galileo, imagination is "not an addition to perception; it is its living vehicle". The limit of vision is the limit of conception.

> Sculpture is not a science but a very mechanical art [...] A sculptor only need know the simple measurement of the limbs and the nature of movements and posture. With this knowledge he can complete his works, demonstrating to the eye whatever it is, and not inherently giving any other cause for admiration in the spectator, unlike painting, which on a flat surface uses *the power of its science* to display the greatest landscapes with their distant horizons.[19]

Through Leonardo's eye, "painting is a second creation made with imagination", whereas for Galileo philosophy never signifies a product of imagination: "La cosa non istà cosi" (Galilei 1623; 1957, p. 121). Leonardo's painting combines the power of imagination with "the power of its science", namely with *perspectiva pingendi*. What are the characters of a second creation made with imagination? How does it convince and enchant us with the strength of a mathematical demonstration?

Already in medieval philosophy, the ideal of a contemplative science of Greek-Arabic tradition, putting itself in dialectic relation with an active conception of a partly Stoic, as well as neo-Platonic, matrix, allowed room for a "doctrine of light" as a demonstrative science based on the geometric optic rules, the *natural* perspective. It seems of interest to observe that, following the theory of rays by Al-Kindi (2003), resumed by medieval *perspectivi*, if light is the "substance" generating all natural things, the rays described by Euclidean geometry had to be "materialized". Carrying over concepts and methods of the medieval perspective into a plane surface, the Renaissance artists invented the "*artificial* perspective". The new *inventio*, however, asked Al-Kindi's rays to turn again into pure forms;

---

[19]"La scultura non è scienza ma arte meccanicissima, perché ... in sé finisce dimostrando all'occhio quel che quello è, e non dà di sé alcuna ammirazione al suo contemplante, come fa la pittura, che in una piana superficie per forza di scienza dimostra le grandissime campagne co' lontani orizzonti" (*Trattato*, I.31).

thus, freed from their substantial character, the rays were able to follow different rules (travelling in parallel they could meet at one point) and to construct a new representation space. The result is a painted or drawn scene, which is supposed to be indistinguishable from the image transmitted by a glass or reflected by a mirror. It is achieved by projecting the three-dimensional scene onto a plane, letting the flight lines converge to a central point specularly symmetrical to the unmoving eye of the painter–observer. Thus every image is anchored to its author–creator in the representation space.

But perspective is not solely a technique of representation. For Alberti, Piero della Francesca, and Leonardo, and even more for Dürer, it is both an "optical art" and a "science of painting".[20] It makes painting a science, rigorously deduced from Euclidean geometry; it makes geometric optics an art, capable of creating a symbolic and imaginary reality different from any model drawn from optical perceptions. It is not an art intended as a pure fiction independent of the constraints of natural necessity; it is not a science intended as an accurate description of natural necessity. It is, however, *costruzione legittima*, for it links artistic representation with the scientific vision of its author: an art of drawing, which binds its freedom of representation to mathematical rules; a science of seeing, which asks the imagination's eye for the principles of description. Generated from art and science, perspective finds its meaning in binding one to the other and its function in making one commensurable with the other.

In his art and his scientific conception, Albrecht Dürer better interpreted this interplay between the art of geometry and the science of drawing that constitutes the essence of perspective. As a reader of Euclid, Apollonius, and Platonic mathematicians, he understood how measure must be at the foundation of drawing and painting. Only in this way, sharing the rigor and the objectivity of geometry, painting can be emancipated from the fortuity of an artisan practice and transform itself from a mechanical to a liberal art. As an artist, Dürer realized the insufficiency of the traditional geometry that was transmitted from *Elements* and shared by mathematicians of the sixteenth century, able to describe profiles of objects but not to regulate the forms and dynamics of living things. For this reason, he asked the "science of measure" to furnish itself with instruments to cope with a dimension of reality (invisible, mutable, lively, magical, cursed, or negative) that escapes not only natural description, but also propositions and theorems of Euclidean geometry.

It is in the solids of Pacioli-Leonardo that Dürer believed to envision a model for a geometry more sophisticated (*subtilis*) than that of Euclid: a "constructive" geometry, which he believed to recognize in the generation of complex and irregular forms obtained by sectioning the angles of Platonic polyhedra. It is an art of geometry that, combining Euclid with Plato, with Apollonius, with the magic and mysterious mathematics of *Melanconia I*, recognizes measure as a common rule

---

[20]See also El-Bizri's essay in this book.

of a reality to be represented and of the art which represents it.[21] Perspective is, therefore, an *art of measure* propaedeutic to the art of drawing and painting, while the geometry founded on measure becomes an inherent part of the painting itself.

The invention of artificial perspective can be viewed as an accomplishment of Alberti's humanism or, in other words, as an expression of a vision of the world "commensurable to man". According to Alberti, perspective is the construction of harmonious proportions within a representation as a function of the distance. All this is measured in relation to the person who observes through an open window. Thus the world becomes commensurable to man, and such that man could construct an adequate representation of his point of view.[22]

Non possumus ludere soli, the condition posed by Cardano on the game of dice is also the condition for perspective painting: the scene (object to be painted) needs to meet the eye of the painter in order to become an "artefact", and therefore, an intermediate space is needed. The painting is ideally located in between, in the symbolic and relational space that is *Alberti's veil*: a window intersecting the visual pyramid and a mirror reflecting the painter's eye. Performing this double function (of transmission and reflection), the veil allows subject and object to be connected and the distance between to be measured.[23]

An eloquent illustration of the non-separability of author-creation, observer-observed, or subject-object at the core of the art of measure, is provided by *The Arnolfini Portrait* by Jan Van Eyck (see Fig. 1.1 in Chap. 1). The Arnolfinis are painted frontally, but a mirror located behind them reflects an image, hidden to the observer watching the scene, back to direct vision (Fig. 6.5). What is striking, beyond the artistic result, is the complete "double-sided" image of the Arnolfinis, simultaneously visible in front and behind. To this first front–back reflection

---

[21] As *Underweysung der Messung* (1525) documents, and as the style of the artist confirms, Dürer conceived a science of measure that obtained the most original results when it applied procedures typical of an artist's workshop (*bottega*) to abstract mathematical objects. Applying the method of double projections, in use by carpenters and architects, to conic sections, Dürer obtained a construction that Gaspard Monge would theoretically codify at the end of the eighteenth century in his "descriptive" geometry.

[22] "First of all, on the surface on which I am going to paint, I draw a rectangle of whatever size I want, which I regard as an open window through which the subject to be painted is seen; and I decide how large I wish the human figures in the painting to be. I divide the height of this man into three parts, which will be proportional to the measure commonly called a 'braccio'; for, as may be seen from the relationship of his limbs, three 'braccia' is just about the average height of a man's body. With this measure I divide the bottom line of my rectangle into as many parts as it will hold; and this bottom line of the rectangle is for me proportional to the nearest transverse equidistant quantity seen on the pavement. Then I establish a point in the rectangle wherever I wish; and as it occupies the place where the centric ray strikes, I shall call this the centric point" (Alberti 1436; 2004, p. 54).

[23] "Perspective is by nature a two-edged sword", Panofsky wrote, because it "subjects artistic phenomenon to stable and even mathematically exact rules", but "the way [these rules] take effect is determined by the freely chosen position of a subjective 'point of view'" (Panofsky 1924–25; 1997, p. 67).

Fig. 6.5 Jan van Eyck: *The Arnolfini Portrait*, 1434. Detail of the mirror

corresponds another one, less obvious, of observer–painter. Approximately at the point where, according to the rules of the linear perspective, the flight lines orthogonal to the plane would converge,[24] we can distinguish the figure of the painter and read the caption: *Johannes de Eyck fuit hic 1434*. "Hic" means on the mirror behind the Arnolfinis, in the point specularly symmetrical to the eye of the painter–observer. The fact that van Eyck does not master the "correct" rules of perspective (the orthogonal lines do not converge in a single vanishing point) urged him to "declare" the artifices used to drive the spectator toward the desired effect. All this constitutes an advantage for the interpretation, in that it reveals, out of any possible pretense, the intentions prefixed to the representation and that it documents a trend which, in the Renaissance, was not exclusive of visual arts.

In this painting, the details of the mirror-image of the painter and the ancillary role of the writing (*Johannes de Eyck fuit hic*) make evident what, some decades later, the "rigorous" linear perspective would make possible: the display of the painting in space, rather than the traditional flattening of the scene within the frame of the picture. To read the painting, the panel must be unfolded in all the directions of space: towards the back, up to the wall which accommodates the mirror, towards the front where one can locate the painter reflected by the mirror, and towards the side where an open window brings back *lontani orizzonti*. In this reading, the reality of the painting, object of sensible experience, becomes a portion of a much wider reality, constructed *by art*. Here, the mirror adds a symmetry plane to the panel and

---

[24]For more details on this construction, see Stillwell's essay (chap. 1).

presents the painter–observer with two "correlated" points of view: one "natural", within the sensible reality of the person who observes the painting and sees the frontal scene; the other "artificial", constructed by art, beyond the plane of the representation, within the "reality" of the mirror which renders the hidden backside of the scene. This "dual" perspectival representation allows the painter to face the *veil* from both sides, to be alternatively observer-subject and observed-object, because the two conditions—observing and being observed—are symmetrical, or mutually transformable, thanks to the action of the mirror. The key role of the mirror in this captivating painting is reminiscent of Alberti's vision of painting as an invention of Narcissus[25]:

> I would venture to assert that what ever beauty there is in things has been derived from painting. *Painting was honoured by our ancestors with the special distinction that, whereas all other artists were called craftsmen, the painter alone was not counted among their number.* Consequently I used to tell my friends that the inventor of painting, according to the poets, was Narcissus, who was turned into a flower; for, as painting is the flower of all the arts, so the tale of Narcissus fits our purpose perfectly. (Alberti 1436; 2004, p. 61)

In the *Arnolfini Portrait* one can already appreciate the widening of the "pictorial space" which would be formalized by central perspective rules. A somehow similar widening of the representation space, triggered by complex numbers, would enable Hilbert space to accommodate quantum theory. Yet the subject–object "correspondence", masterfully captured by van Eyck and inherent to any form of perspectival representation, is spotted by John Bell (1993) as an issue "at the very root of the unease that many people still feel in connection with quantum mechanics":

> it is interesting to speculate on the possibility that a future theory will not be *intrinsically* ambiguous and approximate. Such a theory could not be fundamentally about "measurements", for that would again imply incompleteness of the system and unanalyzed interventions from outside. Rather it should again become possible to say of a system not that such and such may be *observed* to be so but that such and such *be* so. The theory would not be about "*observ*ables" but about "*be*ables". (Bell 1993, p. 41)

But the lesson derived from Renaissance perspective can hardly make sense of Bell's concern. While the variety of forms presented by the painter shows how the pictorial space is determined by the subject, it also shows how the subject gets "entangled" with the object. Indeed, *non possumus ludere soli* does not solely express the natural condition for playing any game, but it also gives voice to a requirement for otherness, relativity, and distinguishability, at the foundation of the concept of measure.

---

[25]For more on the magic of mirrors and symmetries, see Altmann's essay (Chap. 5).

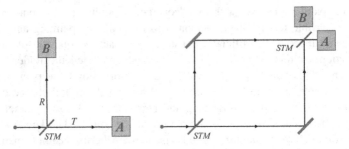

Fig. 6.6 Quantum randomness

6.6 Quantum Events: A Perspectival View

Any physical quantity which can be measured on a physical system constitutes an "observable" of the system. Any physical theory is about observables, but the classical presupposition that observables are made out of objective properties ("beables") of a physical system is not tenable in quantum theory because its observables can be *incompatible*. Incompatible observables of a quantum system are, for instance, position and momentum, spin components, or—as we shall see below—trajectory and interference. Intuitively, an experimental arrangement which allows to answer a yes–no question concerning one of them, prevents to answer a yes–no question concerning the other. More than "incapable of existing together in the same system", incompatible observables are "unable to be held together by the same eye". Nevertheless, Heisenberg's uncertainty principle requires incompatible observables to be *distinguishable* within one and the same representation space. The information about the value of one observable has to be rigorously distinct from the information about the value of another incompatible observable, but the two observables are *bound to each other*.

Consider a photon, a light particle, which encounters a *semi-transparent mirror*. The photon can be reflected (R) or transmitted (T) with the "same probability" (Fig. 6.6, left). This means that if we sent a large numbers of photons, one by one, through the semi-transparent mirror and counted how many of them have taken each route, then we would expect to find that a half of them are transmitted and the rest are reflected. Thus, each photon can be either transmitted or reflected, and *the probability of an event is 1/2*.[26] Two detectors, A and B, can test this probability prediction by measuring (counting) the photons.

Randomness makes betting on the result of one of these measurements as fair as betting on the result of a roll of a coin in a game of heads or tails: in both cases, the ratio between favourable and possible cases is 1/2. Does the comparison between the two games make any sense? How are photons and coins able to behave *at random*?

---

[26]The total over the events must be 1.

In a game of heads or tails, a coin relies upon *its* two faces to behave at random: being neutral in respect of the two alternatives, the coin allows the game to be fair. As (observable) sides *of* a coin, head and tail are taken as "properties" of the coin. Randomness, therefore, rests on interchangeable alternative *properties* of a physical object. A photon, however, needs a semi-transparent mirror to behave at random. It is the semi-transparent mirror that presents the photon with two alternatives; being neutral in respect of them, on one side the photon enables the semi-transparent mirror to *transmit or reflect*, on the other, the semi-transparent mirror enables the photon to be *transmitted or reflected*. "Transmission" and "reflection" are not properties of the photon; we might rather call them "relational events". Before interacting with a semi-transparent mirror, a photon is neither transmitted nor reflected. It is not a photon in isolation that is able to behave at random, but a photon *in relation to* a semi-transparent mirror. Randomness, therefore, is brought about by interchangeable alternative "*correlations*".

Let's double the alternative ways for the photon. By adding two (perfect) mirrors to the original semi-transparent mirror, one on each route (T and R), plus a second semi-transparent mirror, at the meeting point of the two orthogonal rays determined by the mirrors, we get the Mach-Zehnder interferometer shown in Fig. 6.6 (right). Now the photon can reach a measurement (detector A or B) *via* two "symmetrical" alternative paths between the two semi-transparent mirrors. Therefore, there are four alternatives: the photon can be transmitted or reflected twice, transmitted by the first semi-transparent mirror and reflected by the second, and vice versa— TT, RR, TR, and RT. We would expect the photon to be measured (registered) with the same probability by either detector A or detector B.[27] According to the rules of quantum physics, however, if the two paths are exactly symmetrical, i.e., *interchangeable*, then the photon reaches the detector A with certainty (probability 1). The detector B remains "in darkness". Here is evidence of *quantum randomness* and its capability of composing uncertainties in certainty. Here is also "evidence" of *quantum interference* for *one* particle. Indeed, the loss of uncertainty in quantum behavior is usually explained as a consequence of "constructive" interference between the alternative ways in which an event can occur. How can an interference between *possible ways* come about? It seems as if an invisible *quid*, travelling at the speed of light (exactly as a photon does) between the two semi-transparent mirrors, counseled the photon the way to go.[28]

Quantum interference is a master of discretion: put on the spot, it disappears as a creature of shadows. If a measurement reveals the path of the photon between the two semi-transparent mirrors, then the probability that the photon reaches either of the two detectors becomes balanced (1/2 for each). Uncertainty is restored, together with classical probability, and quantum interference vanishes. As mentioned above, path (or trajectory) and quantum interference are incompatible observables: a measurement able to answer the question "which path?" is unable to answer the

---

[27]Because the probability of each alternative is 1/4, the probability of a measurement is again 1/2.

[28]David Deutsch (1997) coined the suggestive term "photon-shadow".

question "which detector?", and vice versa. Notice that a photon in B "informs" that its (intermediate) path has been measured, but does not inform about "which path". Moreover, the probability that the photon has *not* confronted a "measurer" on its way between the two semi-transparent mirrors is 1/2. In other words, a photon can be measured even though no photon-measurer interaction takes place.[29] What is puzzling is not how we get information about "which path",[30] but how a photon travelling along one path can get information about the *possibility* of a measurement along the other. Once more, an invisible *quid* seems in action.

Einstein refused to accept that a theory based on Heisenberg's uncertainty principle could provide a satisfying description of physical world. His most famous attempt to find a flaw was a joint article with Boris Podolsky and Nathan Rosen in 1935: "Can quantum mechanical description of physical reality be considered complete?". According to EPR, physical quantities (observables) which possess definite values must have a counterpart in a *complete* physical theory. Because the uncertainty principle precludes the precise knowledge of two incompatible observables, either quantum theory is not complete or two physical quantities, which the theory predicts with certainty, cannot have simultaneous reality. As a reasonable criterion for reality, they assumed that: "If, without in any way disturbing a system, we can predict with certainty (i.e., with probability equal to unity) the value of a physical quantity, then there exists an element of physical reality corresponding to this physical quantity" (Einstein et al. 1935, p. 777).

In order to show how elements of physical reality are expected to correspond to physical quantities, the EPR argument goes as follows. Suppose to have two particles, A and B, interacting for a while. Once the two systems no longer interact, consider how values can be attributed to their (incompatible) physical quantities, for convenience, say T and R. Notice that quantum theory describes the two particles as a combined system, namely the *pair*. Suppose to perform a measurement of T on the particle A, and to obtain the value T_A. Quantum theory can then predict with certainty the value T_B of the same quantity T of B *at the same time*. Therefore, according to the EPR criterion for reality, there must be an element of physical reality corresponding to T_B. And yet, if a measurement on A had provided the value R_A of the observable R, it would have been possible to predict with certainty the value R_B of the same quantity R of B at the same time. Therefore, there must also be an element of physical reality corresponding to R_B! EPR did not question the impossibility of measuring two incompatible observables on one system, but objected to the conclusion that the reality of the physical quantities of the system B could depend upon the process of measurement carried out on the system A, far away from B. "No reasonable definition of reality could be expected to permit this" (*ibid.*, p. 780).

---

[29]Looking at Fig. 6.6, imagine a detector set on the path T, after the first *STM*. A photon on the path R cannot be measured.

[30]A device which does not detect the photon along T informs that the photon has been reflected.

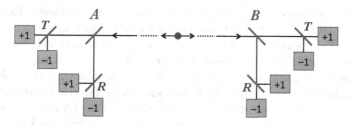

Fig. 6.7 Quantum correlations

In fact, the EPR argument pointed out that quantum theory involves a form of *non-locality* conflicting with the theory of relativity. When a measurement is performed, *instantaneous correlations* between two "separated" physical systems occur. This time, if an invisible *quid* was in action, it should travel faster than light.

To illustrate the point at issue, the experimental arrangement in Fig. 6.7 can be helpful. A source emits pairs of particles. One particle goes to the left, A, the other goes to the right, B. Each particle faces two semi-transparent mirrors on its way to a measurement of one of its "two-value" observables T and R. The two alternative paths, from the source to the measurement of T or R, are *not* interchangeable; hence, quantum interference for one particle does not occur. The probability of an event is 1/4.[31] For each *pair* of particles, however, there are four alternative pairs of measurements: the two particles are both transmitted or reflected, the particle A is transmitted and the particle B is reflected, and vice versa—$T_A T_B$, $R_A R_B$, $T_A R_B$, and $R_A T_B$. All of the measurements are performed at the same distance from the source. The two alternatives $T_A T_B$ and $R_A R_B$—i.e., same observable measured on the left and on the right—are *interchangeable for the pair*. Consistently, quantum theory predicts *with certainty* that the values of the same observable are the same for the two particles. Quantum randomness is also capable of generating perfect correlations.

It is worth stressing that the measurement to be performed is "chosen" *at random* by the two *independent* semi-transparent mirrors labeled A and B in Fig. 6.7.[32] If a pair of detectors reveals the path of each particle between the measurement, i.e. which observable is going to be measured on each particle, the certainty of the correlations is lost. Paths and "correlations" are also *incompatible* observables. Here is evidence of quantum interference for *one pair of particles*.

How can such an inference come about? Reluctant to play with chance, Einstein went in search of "hidden variables". Even though unable to travel faster than light, an invisible *quid* could be set at the source, and give instructions to any pair of particles before departure. Following these instructions, the particles should be able

---

[31]There are four possible results of a measurement.

[32]Therefore, the probability of each alternative pair of measurements is 1/4, and the probability of measuring the same observable is 1/2.

to restrain randomness so much as to produce the correlations. How to fix suitable boundaries for randomness? Bell thought up the following formula: $S = (T_A + R_A)T_B + (T_A - R_A)R_B$. Let the result of each measurement be either $+1$ or -1, as shown in Fig. 6.7, then the value of S can be $+2$ or -2.[33] A large number of measurements on pairs of particles allows the average value of S to be obtained, and the *hidden variables* hypothesis to be tested. Bell's theorem asserts that if the correlations are established at the source, the average value of S must be inside the interval $[-2, +2]$. But quantum theory predicts an average value of S beyond that interval: $2\sqrt{2}$!

Thus, quantum correlations cannot be established at the source, as they violate Bell's inequality.[34] But do they really need an invisible *quid* travelling faster than light? A perspective view might cast light on the issue.

In 1935, Schrödinger already connected quantum correlations with a notion of *non-separability*, called "entanglement". He noted that two quantum systems can interact in a way such that only the properties of the pair are defined. Though any individual system manages to hold a set of well-defined properties, for example, the components of the spin, once two systems get entangled in a pair, the spin of one system *and* that of the other go in the *same* direction or in *opposite* directions. The properties "being the same" or "being opposed" are clearly properties related to two objects. Consequently, quantum theory forges pure "relational properties", which do not work for individual systems.

And yet, how can a measurement on a physical system be defined, if not in relation to a measurer-system? How could a result of a measurement, i.e., an event, have meaning outside the net of the *alternative correlations* between the two systems? Any physical system assembles a number of characteristic "potential" features. All of them need to enter a perspective to be caught by an eye and become (temporarily) "events". Therefore, any quantum event provides evidence of quantum interference for a pair of physical systems, and focuses on an exclusive "relational property" of the pair.[35] To the extent that a quantum measurement is viewed as an interaction, with the twin requirement of freedom in choosing the observable to be questioned and capability of discerning the relevant answer, it calls for sharpening the probability relationships associated with its results and, consequently, for sharpening their mathematical representation.[36] This leads to require the representation space of quantum theory to be *complex*.

---

[33]Each particle can give its observables T and R either the same value or opposite values. As to the instructions of A, they must be either $T_A = R_A$ or $T_A = -R_A$. In the first case, $S = 2T_B$, in the second, $S = 2R_B$. Combining the instructions of A with the instructions of B, the value of S must be $+2$ or -2.

[34]For more details see Scarani (2003).

[35]See Rovelli (1996) for an interpretation of quantum theory which appears in tune with the present view. Cf. also van Fraassen (2008).

[36]For more details, see Wheeler (1990), cf. also Wheeler & Zureck (1983).

In mathematical terms, the conflict between classical probability and quantum interference finds a solution in the notion of a *complex probability amplitude*. Quantum theory defines the probability of an event as the square modulo of its probability amplitude given by a complex number. Since a complex number has a "phase", which measures its angular distance from the real axis in the Argand plane,[37] quantum interference effects can be drawn from complex numbers. When an event can occur in several interchangeable alternative ways, its probability *amplitude* is the sum of the probability amplitudes for each way. In the basic case of two alternatives with probability amplitudes α_1 and α_2, the probability amplitude of the event is $\alpha = \alpha_1 + \alpha_2$, but the probability is not $p = p_1 + p_2 = |\alpha_1|^2 + |\alpha_2|^2$. Indeed, the square modulo of the probability amplitude α gives:

$$p = |\alpha|^2 = |\alpha_1 + \alpha_2|^2 = p_1 + p_2 + \left(a_1^* \alpha_2 + a_1 \alpha_2^*\right).$$

Here, the last term $\left(a_1^* \alpha_2 + a_1 \alpha_2^*\right)$ can be considered responsible for the failure of classical probability or can be praised for quantifying quantum interference. Thus, when there are distinct interchangeable ways in which an event can occur, the probability of the event is the sum of the probabilities for each individual way, refined by an additional term which marks the "angular distance" between any pair of ways. Once a measurement is performed, the alternative taken by the physical system is determined and, therefore, any interference with the others disappears.

Pondering Bell's concern about the intrusion of a measurer in quantum theory, and Einstein's about "a God who plays dice", we might observe that because quantum theory "is fundamentally about the result of 'measurements', and therefore presupposes in addition to the 'system' (or object) a 'measurer' (or subject)" (Bell 1993, p. 40), it does need a *veil* similar to Alberti's one to accommodate those results. Alberti (1436) was the first to discover the usage of a veil in painting:

> It is like this: a veil loosely woven of fine thread, dyed whatever colour you please, divided up by thicker threads into as many parallel square sections as you like, and stretched on a frame. I set this up between the eye and the object to be represented, so that the visual pyramid passes through the loose weave of the veil. (Alberti 1436; 2004, p. 65)

How to use it in quantum physics? The grid of the veil, which separates the rays of a visual pyramid, determines how many values of one observable can be measured. Here is a condition on observability, to be compared with the orthogonality of the rays spanning the Hilbert space of one observable. The orientation of the veil, namely the angular distance between the vertices of alternative visual pyramids (converging on the same physical system), seals off the observable to be focused on. Here is a criterion for distinguishing *incompatible* observables, to be compared

---

[37] The Argand plane is a two-dimensional plane where we can visualize any complex number c as a point and locate it by means of Cartesian coordinates (x, y) such that: $c = (x + iy)$ or in polar form as $c = |c| e^{i\theta} = |c| (\cos\theta + i\sin\theta)$.

with the "obliquity" of the rays related to those observables on the Hilbert space. In this perspectival frame, we may venture to say, the complex character of quantum probability becomes significant. Quantum interference is triggered by the measurer-system correlations through the veil.

As much as knowledge—be it scientific or artistic—grows out of a perspective view, it demands a representation space able to render the symmetrical relation between system-object and observer-subject. This not only carries us to the point where the Euclidean space turns into the perspective space of the Renaissance painting, but it also encourages the eye to see as far as contemporary age, and appreciate the architecture of the complex space where quantum theory depicts its image of physical reality.

References

Alberti, L. B. (1436). In C. Grayson (Ed.), (1975). *De pictura*. Roma-Bari: Laterza [English translation: *On painting*. London: Penguin Books (2004)].

Alberti, L. B. (1485). In G. Orlandi, & P. Portoghesi (Eds.), (1966). *De re aedificatoria*. Milano: Il Polifilo [English translation: *On the art of building*. Cambridge: MIT Press (1988)].

Al-Kindi (2003). *De radiis* (Arabic text IX century; Latin trans. XII century). Paris: Éditions Allia.

Bell, J. S. (1993). *Speakable and unspeakable in quantum mechanics*. Cambridge: Cambridge University Press.

Bombelli, R. (1572). *Algebra*, edited by E. Bortolotti. Milano: Feltrinelli (1929).

Bortolotti, E. (1929). *Preface to Bombelli's algebra*. Milano: Feltrinelli.

Cardano, G. (1544). *De sapientia libri V*, Norimbergae apud Iohannem Petreium. In *Opera Omnia* (Vol. I, New Latin edition by M. Brancali). Florence: Leo S. Olschki (2008).

Cardano, G. (1545). *Artis magnae, sive de regulis algebrici liber unus*. Norimbergae apud Iohannem Petreium. In *Opera Omnia* (vol. III) [English translation: *The rules of algebra*. Cambridge: MIT Press (1968)]. New Latin edition by M. Tamborini. Milano: Franco Angeli (2011).

Cardano, G. (1550). *De subtilitate libri XXI*, Lugduni apud Gulielmum Rouillium sub scuto Veneto. In *Opera Omnia* (Vol. III).

Cardano, G. (1562a). *Neronis encomium*, Basileae ex officina Henric Petri. In *Opera Omnia* (Vol. I).

Cardano, G. (1562b). *Encomium geometriae recitatum anno 1535 in Academia Platina Mediolani*. Basileae ex officina Henric Petri. In *Opera Omnia* (Vol. III).

Cardano, G. (1663a). *Liber de ludo aleae*. In *Opera Omnia* (vol. I) [Engl. trans. in Ore (1953, pp. 181–241)]. New Latin edition by M. Tamborini. Milan: Franco Angeli (2006).

Cardano, G. (1663b). *Opera Omnia, tam hactenus excusa; hic tamen aucta et emendata, quam nunquam alias visa ac primum ex auctoris ipsius autographis eruta* (10 vols.) Lugduni sumptibus Ioannis Antonii Huguetan et Marci Antonii Ravaud.

Cassirer, E. (1927). *The individual and the cosmos in renaissance philosophy*. Philadelphia: University of Pennsylvania Press (1963).

Deutsch, D. (1997). *The fabric of reality*. London: The Penguin Press.

Einstein A., Podolski B., & Rosen N. (1935). Can quantum mechanical description of physical reality be considered complete? *Physical Review 47*, 777–780.

Field, J.V. (1997). *The invention of infinity*. Oxford: Oxford University Press.

Field, J.V. (2005). *Piero della Francesca. A mathematician's art*. New Haven and London: Yale University Press.

Galilei, G. (1623). *Il Saggiatore*. In F. Flora (Ed.). *Opere*. Milano-Napoli: Ricciardi (1957).

Hacking, I. (1975). *The emergence of probability*. Cambridge: Cambridge University Press.

Klein, J. (1968). *Greek mathematical thought and the origin of algebra*. Cambridge, MA: MIT Press.

Kline, M. (1953). *Mathematics in western culture*. New York: Oxford University Press.

Kline, M. (1972). *Mathematical thought from ancient to modern times*. Oxford: Oxford University Press.

Koyré, A. (1958). Les sciences exactes. In R. Taton (ed). *Histoire générale des Sciences* (Vol. 2, pp. 12–51). Paris: Presses Universitaires de France.

Leonardo da Vinci, *Trattato della Pittura*. Codice Vaticano Urbinate 1270. Roma: Unione Cooperativa Editrice (1890).

Ore, Ø. (1953). *Cardano. The gambling scholar*. Princeton: Princeton University Press.

Pacioli, L. (1509). *De Divina Proportione*. Venice: Paganus Paganinus de Brixia. Reprinted by Dominioni, Maslianico (1967).

Panofsky, E. (1924–25). Die Perspektive als 'Symbolische Form'. *Vorträge der Bibliothek Warburg* Leipzig-Berlin: Vorträge. B.G. Teubner [English translation: *Perspective as symbolic form*. New York: Zone Books (1997)].

Piero della Francesca (1474). *De Prospectiva Pingendi*, edited by G. Nicco-Fasola. Florence: Casa Editrice Le Lettere (1984).

Rovelli, C. (1996). Relational quantum mechanics. *International Journal of Theoretical Physics*, *35*, 1637–1678.

Scarani, V. (2003). *Initiation à la Physique Quantique* Paris: Vuibert [English translation: *Quantum physics. A first encounter*. Oxford: Oxford University Press (2006)].

Schrödinger, E. (1935). Die gegenwärtige Situation in der Quantenmechanik. *Naturwissenschaften*, *23*, 807–812.

van Fraassen, B. (2008). *Scientific representation*. Oxford: Oxford University Press.

Weyl, H. (1952). *Symmetry*. Princeton: Princeton University Press.

Wheeler, J. A. (1990). Information, physics, quantum: The search for links. In Zurek W. H. (Ed.), *Complexity, entropy, and the physics of information*. Redwood City: Addison Wesley.

Wheeler, J. A., & Zureck, W. H. (Eds.). (1983). *Quantum theory and measurement*. Princeton: Princeton University Press.

Chapter 7
Radices Sophisticae, Racines Imaginaires: The Origins of Complex Numbers in the Late Renaissance

Veronica Gavagna

The aim of this chapter is to clarify what is meant by the "invention of complex numbers" by the Renaissance Italian algebraists Girolamo Cardano and Rafael Bombelli. It will be demonstrated that, despite the *radix sophistica* found in Cardano's *Ars Magna* that indicates the expressions $a \pm b\sqrt{-1}$, Cardano could not arithmetically operate with them, because he was not able to determine the sign of the square root of a negative number. Vice versa, Bombelli overcame this problem by inventing not "imaginary numbers", but rather the new signs "plus of minus" (*più di meno*) and "minus of minus" (*meno di meno*) and their rules of composition. The *radices sophisticae* of Cardano and Bombelli were thus entities able to give meaning to Tartaglia's solution formula for a cubic equation in the irreducible case, just as the *racines imaginaires* of Albert Girard and René Descartes gave meaning to the first (weak) formulations of the Fundamental Theorem of Algebra. Ultimately, I show that in the late Renaissance the *radice sophisticae* or *racines imaginaires* were something quite different from the modern "complex numbers", essentially because they appeared only as a useful tool to solve problems, and not yet as a true mathematical object to be studied.

7.1 Girolamo Cardano and His *Ars Magna*

The *radices sophisticae*, expressions of the form $a \pm b\sqrt{-1}$ that might be regarded by a modern reader as complex numbers, appear in the chapter of Girolamo Cardanus's *Ars magna* (1545) devoted to investigating the existence of *false* roots of a quadratic equation. Since the only admitted *true* roots were natural numbers, positive fractions and radicals, the term *false* for Cardano denoted, in modern

V. Gavagna (✉)
Department of Mathematics, University of Salerno, Via Giovanni Paolo II, 132 - 84084 - Fisciano (SA) - Italy
e-mail: vgavagna@unisa.it

R. Lupacchini and A. Angelini (eds.), *The Art of Science*,
DOI 10.1007/978-3-319-02111-9_7,
© Springer International Publishing Switzerland 2014

language, both negative and complex roots. Concerning the meaning to be attributed to these strange mathematical objects, the author concludes that neither of the two kinds of roots admits a geometric representation, yet in the case of negative solutions it is sometimes possible to give a mercantile interpretation, imagining that they could represent a debt. Although Cardano allowed himself to neglect the *radices sophisticae* in this context, by assuming that a quadratic equation with negative discriminant is not actually solvable, he must have considered the problem when the *radices* appear in the solution formula of certain cubic equations, since the three solutions of the equation are real. In this case it is not possible to ignore the *radices sophisticae*.

Before getting to the heart of the matter, it is worth outlining the historical-mathematical context where the algorithm for solving cubic equations, the first important original result of modern mathematics with respect to the classical tradition, came to maturity.

It is well known that the solution formula of the third degree equations was published in the *Ars magna*,[1] but it is probably less known that Cardano began dealing with this problem a few years before, when he started the redaction of his first printed mathematical work, the *Practica arithmetice et mensurandi singularis*.

Started in 1537 and published in 1539, the *Practica* is a treatise on arithmetic and practical geometry, enriched with a section devoted to algebra that places the work in the medium-high range of the so-called abacus treatise. The *Practica* is characterized by a continuous comparison to the *Summa de arithmetica, geometria proportioni et proportionalita* (1494) by Luca Pacioli: this is not surprising because anyone who experimented with this genre in the first decades of the sixteenth century, could not ignore the *Summa*, an inescapable point of reference for the abacus world. In order to be able to compete with such a ponderous encyclopedia of arithmetic and practical geometry interspersed by long theoretical digressions, Cardano decided to write a handbook with nearly antithetical features, thus drafting an easy and enjoyable text, divided into two main parts: the first consisting of concise calculation rules accompanied by few explanatory examples; the second providing additional exercises and applications in a mix of arithmetic and geometric problems. To emphasize the distance from the *Summa*, Cardano devoted the initial pages of the *Practica* to the enumeration of the novelties contained in his own work interspersed with stinging criticisms against Pacioli and the best known abacus authors of the time, like Giovanni Sfortunati and Pietro Borghi. These criticisms, generally well-founded, culminated in the final chapter of the *Practica*, a meticulous listing and analysis of the errors contained in the *Summa*. While these devices were evidently intended to carve out a space in the national book market, the choice of writing the treatise in Latin instead of vernacular Italian, common for

---

[1] After the *editio princeps* printed in Nuremberg in 1545, Cardano published a new edition of the *Ars magna* in 1570, the same version (filled by many misprints) published in the fourth volume of the *Opera omnia*, edited in 1663 by Charles Spon. For the references, we have used, quite freely, the English translations by T. R. Witmer Cardanus (1968).

this literary genre, made the *Practica* one of the main means of spreading Italian abacus mathematics throughout Europe.[2]

While he was writing the *Practica*, Cardano learned that the mathematician Niccolò Tartaglia from Brescia (1499–1557) had found the solution formula for cubic equations and used it to win a public challenge against Antonio Maria Fior. Such a discovery was very important not only for its intrinsic mathematical importance, but also because it could be used to score another point against Pacioli, who was skeptical about the possibility of finding a general solution formula for cubic equations, since nobody had gone beyond the resolution of a particular case.[3] Achieving such a formula became an important aim for Cardano. Acting with great persistence and applying some clever tricks, he succeeded in extorting it from Tartaglia, though in "encrypted" form and constrained by the promise of not reporting it in his forthcoming *Practica*.[4]

In the *Practica*, the chapters dedicated to algebra follow a well-established pattern of presentation, which starts from binomial equations or equations reducible to them, then continues with quadratic, biquadratic and trinomial equations. Honouring his pledge, Cardano did not consider the general case of cubic equations, but only treated some special cases solved through particular devices. For instance, given the equation $x^3 = bx + c$ with $b, c > 0$, some specific cases were discussed, where the right-hand side could be reduced to particular forms $c = b - 1; c = 2b - 8; c = 1 - b$, in such a way that the sums $x^3 \pm 1$ or $x^3 \pm 8$ could be decomposed as a product of a linear binomial multiplied by a second degree trinomial, thus reducing the degree of the equation to be solved.

It is worth observing that we are adopting here a modern formalism, very far from the algebraic language of the Renaissance, which was essentially rhetorical and therefore unfamiliar to a modern reader. We will adopt current symbolism but, to avoid betraying the original spirit, we will not refer, when speaking of cubic equations, to the general form $ax^3 + bx^2 + cx + d = 0$, preferring to maintain the traditional classification into three canonical cases or *capitula*, as a consequence of the fact that the coefficients p and q had to be positive numbers:

1. $x^3 + qx = p$, "cubus et res aequalia numeris"
2. $x^3 + p = qx$, "cubus et numerus aequalia rebus"
3. $x^3 = p + qx$, "cubus aequalis numeris et rebus"

---

[2]On the *Practica Arithmetica* and the abacus tradition, see Gavagna (2010).

[3]In fact, he wrote in the *Summa*: "But of number, thing and cube together being composed ... it was not possible to find general rules ... except sometimes gropingly for some particular cases ... the art has not yet shown them as there are no ways to square the circle" ("Ma de numero, cosa e cubo tra loro stando composti ... non se possuto finora troppo bene trovar regole generali ... se non ale volte a tastoni in qualche caso particulare ... larte ancora a tal caso non a dato modo si commo ancora non e dato modo al quadrare del cerchio" (Pacioli 1994, c. 150r)).

[4]The history of the solution formula for cubic equations and the challenge between Cardano and Tartaglia before, and between Ludovico Ferrari and Tartaglia next, is one of the best known in the history of mathematics and for this reason we will not enter into details. However, for a slightly different reconstruction, based on a new reading of the extant documents, see Gavagna (2012).

In the best abacus environments, it was probably known that the lack of the quadratic term did not undermine the generality of the equations, since a complete cubic equation could always be reduced, by a simple linear transformation, to a cubic equation without the quadratic term.[5]

In the *Practica*, Cardano did not point out these general cases but, as we have said, he merely treated some special cases solvable with tricks, although he could possibly have had the well-known rhyme by Tartaglia, whose verses concealed the solution formula of the cubic equation.[6] As one can see from the modern "translation" placed between square brackets, this rhyme actually described the algorithm step by step,[7]

> When the cube and things together
> Are equal to some number, $[x^3 + px = q]$
> Find two other numbers differing in this one $[u - v = q]$
> That their product should always be equal
> Exactly to the cube of a third of the things. $[uv = (p/3)^3]$
> The whole remainder
> Of their cube roots subtracted
> Will be equal to your principal thing $[x = \sqrt[3]{u} - \sqrt[3]{v}]$

Thus, in the first case $x^3 + px = q$, one has to solve the system of two equations in the unknowns u and v

$$\begin{cases} u - v = q \\ uv = \frac{p^3}{27} \end{cases} \tag{7.1}$$

which is transformed into the quadratic resolvent

---

[5] The manuscripts *Fond.Princ.II.V.152* and *Conv.Sop.G.7.1137* kept by the National Library of Florence, presumably written in Florence in the last decade of the fourteenth century, have preserved some examples in which cubic equations without a linear term are transformed by a linear replacement, to cubic equations without a quadratic term, which are then solved gropingly ("a tastoni"). Nothing is known about the distribution of such results in the abacus environments. On this question, see Franci (1985). For a transcription of the algebraic section of the ms. *Fond.Princ.II.V.152* we refer to Franci and Pancanti (1988).

[6] The device of the rhyme is less bizarre than it could seem at first sight. Tartaglia was an abacus teacher and it was a common practice to expect from the students the memorization of the most important procedures via acronyms or rhyming verses. Cardano himself, in Chapter V of the *Ars Magna*, offers three *carmina* for solving quadratic equations: "Querna da bis, Nuquer admi, Requan minue dami" (Cardanus 1545, ff. 10v–11v). With regards to mathematics and poetry in the Renaissance, see Saiber (2014).

[7] "Quando chel cubo con le cose appresso/Se agguaglia a qualche numero discreto/Trovan due altri differenti in esso/Ch'el lor produtto sempre sia eguale/Al terzo cubo delle cose netto/El residuo poi suo generale/Delli lor lati cubi ben sottratti/Varra la tua cosa principale" (Tartaglia 1546, f. 123r.).

$$t^2 - qt - \frac{p^3}{27} = 0 \tag{7.2}$$

and has solutions

$$u = \frac{q}{2} + \sqrt{\frac{q^2}{4} + \frac{p^3}{27}} \tag{7.3}$$

$$-v = \frac{q}{2} - \sqrt{\frac{q^2}{4} + \frac{p^3}{27}} \tag{7.4}$$

Since

$$x = \sqrt[3]{u} - \sqrt[3]{v} \tag{7.5}$$

it follows that

$$x = \sqrt[3]{\frac{q}{2} + \sqrt{\frac{q^2}{4} + \frac{p^3}{27}}} - \sqrt[3]{\sqrt{\frac{q^2}{4} + \frac{p^3}{27}} - \frac{q}{2}} \tag{7.6}$$

Tartaglia next considers the second case[8]

> In the second of these acts,
> When the cube remains alone, $[x^3 = px + q]$
> You will observe these other agreements:
> You will at once divide the number into two parts $[u + v = q]$
> So that the one times the other produces clearly
> The cube of the third of the things exactly. $[uv = (p/3)^3]$
> Then of these two parts, as a habitual rule,
> You will take the cube roots added together,
> And this sum will be your thought $[x = \sqrt[3]{u} + \sqrt[3]{v}]$

After carrying out the required computations, the unknown assumes the form

$$x = \sqrt[3]{\frac{q}{2} + \sqrt{\frac{q^2}{4} - \frac{p^3}{27}}} + \sqrt[3]{\frac{q}{2} - \sqrt{\frac{q^2}{4} - \frac{p^3}{27}}}. \tag{7.7}$$

---

[8]"In el secondo de cotesti atti/Quando che'l cubo restasse lui solo/Tu osservarai quest'altri contratti/Del numero farai due tal part'à volo/Che l'una in l'altra si produca schietto/El terzo cubo delle cose in stolo/Delle qual poi, per comun precetto/Terrai li lati cubi insieme gionti/Et cotal somma sara il tuo concetto" (Tartaglia 1546, ff. 123r–123v).

Finally, Tartaglia ends by observing that the last case $x^3 + q = px$ depends on the previous one $x^3 = px + q$, because it has the same roots, but with opposite signs.[9]

> The third of these calculations of ours
> Is solved with the second if you take good care,
> As in their nature they are almost matched.
> With quick steps and light feet these solutions I found
> In one thousand five hundred thirty and four
> My foundations quite sure and certainly sound
> In the watery city surrounded by shore.[10]

After having correctly interpreted the rhyme and tested the goodness of the algorithm, Cardano realized that in the second case $x^3 + px = q$ ("cubus aequalis numeris et rebus") and the third related one, the solution formula did not work when the cube of the third part of the coefficient of the unknown is greater than the square of the half part of the constant term ($\frac{p^3}{27} > \frac{q^2}{4}$). In this case, the square root of a negative number must be computed, and since this operation is not possible, the unknown is not "reducible" to a difference of cube roots; hence, the frequently used expression "irreducible case".

Cardano immediately wrote to Tartaglia to obtain clarification, but the latter did not understand (or perhaps pretended to misunderstand) the legitimate question and answered by accusing the former of not rightly understanding his solution.[11] Cardano had actually fully understood the weakness of the algorithm and was totally aware that he could not ignore it easily, since numerous examples suggested the existence of three roots (real and distinct). Moreover, it was not even a case to be kept on the boundary of an exhaustive treatment of cubic equations because, as he would show in the *Ars Magna*, many types of complete cubic equation, or those free from a linear term, could be transformed into an equation of this type.

---

[9]Tartaglia was in fact completely aware that the sum of the roots (with opposite sign) and their product are respectively equal to the coefficients of the linear term and the constant term.

[10]"El terzo poi de questi nostri conti/Se solve col secondo se ben guardi/Che per natura son quasi congionti/Questi trovai, et non con passi tardi/Nel mille cinquecenté quatro e trenta/Con fondamenti ben sald'é gagliardi/Nella citta dal mare intorno centa". I would like to thank Arielle Saiber for providing me with this translation, that will appear in Saiber (2014).

[11]On August 4, 1539 Cardano wrote to Tartaglia: "I have asked you for the answer to several questions you have never answered, e.g. the one on the cube equal to things and number ... when the cube of the third part of the things exceeds the square of the half of the number, then I cannot make them follow the equation as it appears" ("io ve ho mandato a domandare la resolutione de diversi quesiti alli quali non mi haveti risposto, et tra li altri quello di cubo equale a cose e numero ... quando che il cubo della terza parte delle cose eccede il quadrato della mita del numero, allora non posso farli seguir la equatione come appare" (Tartaglia 1546, ff.125v–126r)); on August 7 Tartaglia replied "And therefore I reply, and say, that you have not used a good method for solving such a case; also I say that such proceeding of yours is entirely false" ("E pertanto ve rispondo, et dico che voi non haveti appresa la buona via per risolvere tal capitolo; anci dico che tal vostro procedere è in tutto falso" (Tartaglia 1546, ff.126r–127r)).

In order to obtain a general formulation, suited to solving any third or fourth degree equation,[12] the anomalous case had to be solved. The two possible directions along which attempts could be steered—that is, to establish rules to manipulate the *radices sophisticae* or to find a solution formula in which they would not appear— were not a priori mutually exclusive, but became so in Cardano's thinking.

Given the difficult path which led to Tartaglia's formula, it is reasonable to assume that Cardano, at least at first, concentrated his efforts on the first of the two directions, and in fact in the *Ars magna* the evidence of an attempt to manipulate expressions of the form $a \pm \sqrt{-b}$ algebraically can be seen, not in the context of the discussion of the irreducible case but, as already remarked, in the analysis of *false* solutions of quadratic equations. Chapter XXXVII, *On the rule for postulating a negative* (*De regula falsum ponendi*), explores three rules for solving problems with *false* solutions, a term indicating negative solutions ("minus puro"), solutions in which square roots of negative numbers appear ("minus sophistico") and hybrid solutions ("componitur haec regula quasi ex ambobus"), respectively.

The first rule explains how to determine the negative solutions of a quadratic equation of the form

$$x^2 = ax + b \qquad (7.8)$$

("censi uguali a cose e numeri" $a, b > 0$).[13] In this case, the solution formula (expressed in modern symbols) provides the only positive or *true* solution

$$x = \frac{a}{2} + \sqrt{\left(\frac{a}{2}\right)^2 + b} \qquad (7.9)$$

In order to find also the *false* one, Cardano does not modify the solution algorithm of (7.8) in order to compute also negative roots, but keeps it unchanged, preferring instead to exploit the relationships existing between roots and coefficients of quadratic equations. First of all, he suggests solving the "twin" equation

$$x^2 + ax = b \qquad (7.10)$$

which has, with respect to (7.8), roots equal in absolute value but opposite in sign. The *true* solution of (7.10),

$$x = \sqrt{\left(\frac{a}{2}\right)^2 + b} - \frac{a}{2} \qquad (7.11)$$

---

[12]In the *Ars Magna*, Cardano also displayed out the solution formula of fourth degree equations, giving the credit to his pupil Ludovico Ferrari. Ferrari's procedure reduced the solution of the equation to that of a third degree resolvent and it is therefore evident that in this context the irreducible case had to be managed.

[13]It is one of the three canonical quadratic equations $x^2 = ax + b$, $x^2 + ax = b$ and $x^2 + b = ax$ where $a, b > 0$ whose solution formulas, even if in rhetorical form, only provided positive roots.

with opposite sign, becomes the *false* solution of equation (7.8). The numerical examples that support this rule are intended to justify the eventual utility of introducing similar *false* solutions, which, as previously said, although they do not find an adequate interpretation in the context of Euclidean geometry, may be effectively represented in a merchant context—for instance, by debts.

The second kind of *false* solution contains the square root of negative terms, that is the *minus sophistico*. The author begins with an example that might sound like one of the most common problems in abacus arithmetic: "Divide 10 into two parts, such that their product is 30 or 40". The problem gives rise to the equation $x^2 + 40 = 10x$, but—as Cardano continues—"it is clear that this case is impossible, nevertheless, we will work thus" ("manifestum est quod casus seu quaestio est impossibilis, sic tamen operabimur"). If the solution algorithm is formally applied, it yields the two expressions ("partes") $5 + \sqrt{-15}$ and $5 - \sqrt{-15}$, which provide the solution to the problem, as the author underlines, because their sum is 10 and their product, computed by appropriate rules for the multiplication of residuals[14] is exactly 40. The rule is followed by an attempt at geometric representation of the solution formula of $x^2 + 40 = 10x$ which is, however, reduced to the possibility of subtracting a rectangle of sides 4 and 10 from a square of side 5, which is not admissible, because negative areas are meaningless.

While in the "minus puro" case it was sufficient to consider the "twin" equation to compute the *false* solution, here such a strategy is no longer effective. The critical point of the "minus sophistico" is the unavoidable negative sign in the discriminant of the equation; it is exactly the impossibility of deleting the minus sign inside the square root that makes, according to Cardano, the square root of a negative number "as subtle as useless" ("adeo est subtile, ut sit inutile").

We do not have to hastily conclude that the connotation of *arithmetic subtlety* ("arithmetica subtilitas") is negative in Cardano's mathematical thinking; it rather describes an intrinsically interesting object from the mathematical point of view, but one that is not really useful from a practical point of view. In *De subtilitate*, a work devoted to the topic of *subtilitas* and many times reworked and edited,[15] among many mathematical *subtilitates* there are some very interesting results, such as the rewriting of proofs of the propositions of the *Elements* based on the use of a ruler and a compass with fixed rather than variable opening.[16]

---

[14]The terms *binomial* (*binomium*) and *residual* (*recisum*) or *apotome* are taken from the Latin translations of Book X of the *Elements*, devoted to the classification of the quadratic irrationals. They respectively denote expressions of the form $a + \sqrt{b}$ and $a - \sqrt{b}$ or, more generally, $a + x$ and $a - x$, where the terms involved have different natures ("quantitas quae additur vel detrahitur, non est eiusdem naturae cum prima").

[15]The three most important editions are dated back to 1550, 1554 and 1560. For a critical edition of the first seven books, compare Nenci (2004).

[16]The chapter *It is shown how any proposition of Euclid's Elements can be proved without a change of opening of the compass (Quomodo quaecumque in Elementis Euclidis demonstrata sunt absque ulla propositi unus tantum circuli mutatione ostendi possint)* from Book XV of the *De subtilitate* essentially represents the Latin translation of the answer to Tartaglia, published by

Although Cardano proposes this result as a rather childish boast void of a real utility ("ostentatione potius iuvenili quam utilitate manifesta"), it is a mathematical result worthy of respect; to have a more general result, we have to wait for *La geometria del compasso* (1797) by Lorenzo Mascheroni and the nineteenth century studies by Jean Victor Poncelet and Jacob Steiner.[17]

The third and last rule of Chap. XXXVII of the *Ars Magna* regards another kind of "minus", which can be considered, according to Cardano, as a combination of the previous ones. The rule is illustrated only by an example, without any comment. It deals with "finding three proportional numbers the square root of the first of which, subtracted from the first, gives the second and the square root of the second, subtracted from the second, gives the third". By supposing that the first quantity is denoted by x^2, the proportion may be written as

$$x^2 : (x^2 - x) = (x^2 - x) : (x^2 - x - \sqrt{x^2 - x}) \qquad (7.12)$$

Cardano finds as solutions the three numbers $\frac{1}{4}, -\frac{1}{4}, -\frac{1}{4} - \sqrt{-\frac{1}{4}}$.

From his point of view, the square of the second term, namely $\frac{1}{16}$, is equal to the product of the first times the third, because

$$\frac{1}{4} \cdot \left(-\frac{1}{4} - \sqrt{-\frac{1}{4}} \right) = -\frac{1}{16} + \frac{1}{8} = \frac{1}{16} \qquad (7.13)$$

The result therefore presupposes the identity

$$-\frac{1}{4} \cdot \sqrt{-\frac{1}{4}} = \sqrt{\frac{1}{64}} = \frac{1}{8} \qquad (7.14)$$

which is based, as it seems, on a hasty attempt to apply the rule of signs "minus by minus equal plus", leading to a completely wrong result.

In actual fact, the critical aspect of these *radices sophisticae* is not really of foundational character but rather of operational nature. Cardano does not question

---

Ludovico Ferrari in the Fifth of the *Cartelli di matematica disfida* Masotti (1974). In fact, Tartaglia challenged Cardano and Ferrari to prove some Euclidean propositions by using, in addition to the ruler, a fixed opening compass, by replacing the third Euclidean Postulate, which allows one to describe a circle with any centre and any distance, with the possibility of describing a circle with any centre, but fixed radius. In the *Quinto Cartello*, Ferrari claimed to be able to prove *all* the Euclidean propositions with a fixed aperture compass and not only those indicated by Tartaglia. In the *De subtilitate*, Cardano recounted that he had re-proved all the Euclidean *Elementa* with Ferrari in very few days ("paucis in diebus"), by using a ruler and a compass with fixed aperture, but he did not mention any of the *querelle* with Tartaglia.

[17]The first, in fact, proved that "any geometric construction that can be performed by a compass and straightedge can be performed by a compass alone" (Mascheroni Theorem), while the latter came to demonstrate that "all Euclidean geometric constructions can be carried out with a straightedge alone, if given a single circle and its centre in addition" (this result is known as "Poncelet-Steiner Theorem" and was definitely proved in 1833).

whether these new objects are numbers or not because, for the mathematicians of his time, number was defined at the beginning of Book VII of *Elements*: "A number is a multitude composed of units". However, there are quantities behaving *as if* they were numbers, that is they obey the rules of addition, subtraction, multiplication, division and square roots.[18] In the *incipit* of the *Practica arithmetice* Cardano explicitly declares that "the object of arithmetic is the integer number so, by analogy, there are four objects, i.e. the integer number, such as 3, the fractional number, like $\frac{3}{7}$, the surd number, such as $\sqrt{7}$, and the named number, such as 3 *census*" ($3x^2$, in modern terms).[19] Note that after this isolated methodological statement—the term "analogy", in fact, is used only this once in the *Practica*—Cardano has no longer any hesitation in referring to these quantities as "numeri surdi, fracti, denominati" and dedicates the following chapters to defining the operations between these "numbers" and their properties.

Cardano therefore tries to understand whether the square roots of negative numbers behave "by analogy" like numbers, and the first step to be carried out is to establish whether they are positive or negative quantities. Certainly, for the *radices sophisticae* the usual traditional rule of signs cannot hold because, for instance, since necessarily $(\sqrt{-15})^2 = -15$, we would have the paradoxical existence of a negative square. In order to preserve the rule of signs, it must be admitted that the "sophistic quantity" is not a negative or positive quantity but, as Cardano already stated in the *Ars magna arithmeticae*, "some recondite third sort of thing".[20]

In the *Ars magna*, the problem of the sign of the "sophistic quantity" remains confined to these two examples of Chap. XXXVII and is not developed further; the author will come back to the topic in *De regula aliza libellus*, a short work published in 1570 in one volume together with the *Opus novum de proportionibus* and the second edition of the *Ars magna*. The term *aliza* means "unsolved" and alludes to the irreducible case of the cubic equation[21]: in this work, in fact, Cardano collects the various (and vain) attempts, developed over 30 years, to solve the problem.

---

[18]On the topic, see, in particular Malet (2006).

[19]"Subiectum Arithmeticae numerus est integer, *per analogiam* quatuor subiecta sunt: videlicet numerus integer ut 3, fractus ut $\frac{3}{7}$, surdus ut Radix 7, denominatus ut census tres, quae omnia explicabo" (Cardanus (1539), Caput primum, *De subiectis arithmetice*, Italics mine).

[20]The only printed edition of the *Ars magna arithmeticae* is the one contained in the fourth volume of the *Opera omnia* published by Charles Spon in Lyon in 1633. On the role of the work in the development of Cardano's mathematics, see Gavagna (2012). The work consists of 40 chapters and 40 problems; the 38th problem also deals with the equation $x^2 + 16 = 6x$ and, concerning the negative discriminant, he observes: "Note that $\sqrt{9}$ is either $+3$ or -3, for a plus or a minus times a minus yields a plus. Therefore $\sqrt{-9}$ is neither $+3$ nor -3 but is some recondite third sort of thing" ("Et nota quod R. p̄ 9 est 3 p̄ vel 3 m̄ nam p̄ & m̄ in m̄ faciunt p̄. Igitur R. m̄ 9 non est p̄ 3 nec m̄ sed quaedam tertia natura abscondita" (Cardanus 1663a, p. 373)).

[21]*Aliza*, or *aluza*, is a mispronunciation based on Byzantine pronunciation αλιϫϳα from the Greek ἀ-λυθεῖα, composed by privative α and the aorist passive singular feminine participle of λύω, loose, solve. I wish to thank Paolo d'Alessandro, who kindly provided this information to me. Up until now, the most complete studies of the *De regula aliza* are due to Cossali (1996) and Confalonieri (2013). For an *excursus* of the various approaches to the irreducible case, also see Gatto (1992).

In the *De regula aliza* the *radices sophisticae* are completely left out, but an echo of the problem of the sign survives in Chap. XXII *On the contemplation of plus and minus, and that minus by minus makes minus (De contemplatione $\tilde{p}$ et $\tilde{m}$ et quod $\tilde{m}$ in $\tilde{m}$ facit $\tilde{m}$)*, where Cardano tries to found new sign rules on a geometrical basis.[22] Cardano compares the usual arithmetical rules to squaring binomials and residuals of the form $a \pm x$ (where a and x are not of "the same nature"), with a possible geometrical interpretation. Although his main aim is to prove that "minus by minus gives minus", Cardano is also compelled to justify, invoking a sort of compensation, the reason why apparently "minus by minus gives plus".[23] In the binomial case, the geometric interpretation of the square $(a + x)^2 = a^2 + x^2 + 2ax$ is based on Proposition II.4 of the *Elements*[24] and reduces to a simple completion of the square having side a with a gnomon of side $x + a$: therefore, there is a perfect agreement between the arithmetic rule and the geometric interpretation. In order to square a recise $a - x$, according to Cardano we have to use the Euclidean Proposition II.7[25] and observe that the square of side $(a-x)$ is obtained by subtracting from the square of side a, the square of side x and the two rectangles of sides x and $a - x$, that is $(a-x)^2 = a^2 - x^2 - 2x(a-x)$. Such an identity would prove that the square of side $-x$ is $-x^2$; the positive sign of x^2 that appears in the identity $(a-x)^2 = a^2 + x^2 - 2ax$ is due to the fact that the term $-2ax$ represents the two rectangles of sides a and x (i.e., the gnomon of sides a and x), where the square x^2 is considered twice. This means, as again Cardano states, that to re-establish the equality we need to add another square.[26]

Although Cardano "proves" that minus by minus is equal to minus, he fails to construct an arithmetic of square roots of negative numbers based on this new law of composition. Then, he abandons the topic and handles the (failed) attempt to solve the irreducible case in another way, that is by trying to generalize cases that can be solved by special tricks.

---

[22]For a detailed analysis of this topic, see Tanner (1980).

[23]"Et ideo patet communis error dicentium, quod $\tilde{m}$ in $\tilde{m}$ producit $\tilde{p}$ neque enim magis $\tilde{m}$ in $\tilde{m}$ producit $\tilde{p}$ neque enim magis $\tilde{m}$ in $\tilde{m}$ producit $\tilde{p}$ quam $\tilde{p}$ in $\tilde{p}$ producat $\tilde{m}$. Et quia nos ubique diximus contrarium, ideo docebo causam huius, quare in operatione $\tilde{m}$ in $\tilde{m}$ videatur producere $\tilde{p}$ et quomodo debeat intelligi" (Cardanus 1570a, p. 44).

[24]"If a straight line is cut at random, the square on the whole equals the squares on the segments plus twice the rectangle contained by the segments".

[25]"If a straight line is cut at random, then the sum of the square on the whole and that on one of the segments equals twice the rectangle contained by the whole and the said segment plus the square on the remaining segment".

[26]"Ideo in recisis necesse est operari per septimam propositionem secundi Euclidis loco quartae: & ita quia in illa includitur additio illa quadrati $\tilde{m}$ in multiplicatione unius in partis integrae, in partem dectractam bis supra gnomonem, ideo oportet addere ad $\tilde{p}$ quantum est quadratum partis illius quae est $\tilde{m}$. Ideo ut in binomiis operamur per quartam propositionem, & secundum substantiam quantitatis compositae, ita etiam in recisis quo ad substantiam & vere operamur cum eadem: sed ad nominum cognitionem operamur in virtute septima eiusdem" (Cardanus 1570a, p. 400).

7.2   Bombelli and His *Algebra*

Two years after the publication of *De regula aliza*, Rafael Bombelli published the first three books of his *L'algebra* (1572). The biography of the engineer Bombelli is still largely unknown. Around 1550, during the interruption of the drainage works in the Chiana Valley in Italy, he devoted himself to the drafting of the first edition of his work in Italian vernacular.

After 1567 he spent some time in Rome and here, together with Anton Maria Pazzi, he translated the first five books of Diophantus' *Arithmetica*, but he could not complete his work because of more pressing commitments.[27] The *Arithmetica* had a deep influence on Bombelli, who, after having revised his own work, preferred to print only the first three of the five books of the *Algebra*, reserving the right to publish the remaining two after a complete overhaul. The project was left unfinished because of Bombelli's sudden death. Only in 1929, the historian of mathematics Ettore Bortolotti found in Bologna two handwritten redactions—one of all five books and one limited to the last two—and finally put the entire work at the disposal of scholars.[28]

In the preface of the printed edition of the *Algebra*, Bombelli states that the purpose of the work is not to reveal new discoveries in the algebraic context, but to reduce "to perfect order" a growing discipline that cannot yet rely on good texts "either because of the difficulty of the topic or because of the confused way of writing of the authors". Bombelli thus presents his work as a re-arrangement of largely existing material; the mention of the obscurity of the texts is probably addressed, at least in large part, to the *Ars magna* and the *De regula aliza*, works with which he creates, as we will see, a close dialogue at a distance.

Just as Cardano did in his *Practica*, Bombelli did in the context of Euclidean arithmetic of natural numbers. When he has to operate concretely, however, he does not hesitate to treat as numbers those objects that are not formally numbers, but which behave as numbers.[29] Thus, square and cube roots, as the author points out, are not numbers in a strict sense, but *sides*, respectively square and cubic, of numbers.[30] In the first part of the *Algebra*, Bombelli tries to find the conditions

---

[27]It is worth mentioning that the first printed edition of Diophantus' *Arithmetica* was published in 1575 in Basel by Xylander.

[28]The most recent edition is Bortolotti and Forti (1966). The drafting of Books IV and V found by Bortolotti, however, shows a text still imperfect, certainly not ready for publication.

[29]In the first chapter *Diffinitione del numero quadrato*, Bombelli indirectly evokes Euclid, explaining that "even if the unit is not a number, in the operations it is useful like numbers" ("se bene l'unità non è numero, pur nelle operationi serve come li numeri"). On the concept of numbers according to Bombelli, also see Wagner (2010).

[30]For example, the square root is defined in this way: "The square root is the side of a non-square number; it is impossible to be denominated: however, it is denoted as surd Root or indiscreet, as it would be if one has to take the square of side 20, which does not mean anything else than finding a number that, multiplied by itself, would give 20; which is impossible to be found, 20 being a non-square number" ("La Radice quadrata è il lato di un numero non quadrato; il quale è impossibile

that make the addition and subtraction closed between roots with the same index. Bombelli does not diverge too much with the best treatises of practical arithmetic, but when he describes some geometric constructions that represent square and cube roots of given segments he is more original, supporting them with the most usual algorithms for their computation.[31]

With regards to the extraction of cube roots, Bombelli emphasizes that the problem of finding the *cubic side* of a known segment *l* can be related to the classical problem of the duplication of the cube "much considered by the ancient scholars during Platonic times" ("dalli antichi molto cercato al tempo di Platone") and is equivalent to inserting two mean proportional segments between *l* and a unit segment ("common measure"). Bombelli proposes two constructions "in lines" ("in linea"), which correspond to those attributed to Heron and Plato, found respectively in the *Mathematical Collections* of Pappus, and in the comment of Eutocius to the second Book of Archimedes' *Sphere and cylinder*.[32] Note that Bombelli's interpretation of Plato's geometric construction requires the use of those "material sliding squares", which, as we will see, will also allow the "in lines" representation of the solution of a cubic. In this case we have to trace the unit segment *cd* perpendicular to the given segment *de*, of which the *cubic side* must be found, and place the first sliding square in such a way that one of the sides passes through the point *c*, and the vertex lie on to the prolongation of *de*, and the second sliding square placed so that one side passes through *e* and the vertex lies on the prolongation of *cd*. In this way, two right-angled triangles may be obtained and, due to the corollary to Proposition VI.8[33] the segments *fd* and *dg* are mean proportionals between *cd* and *de* and the segment *fd* is the *cubic side* of *de* (Fig. 7.1).

The discussion gets to the heart of the matter regarding the arithmetic of binomials and residuals, into which Bombelli inserts the chapter *Demonstration that minus by minus should produce plus (Dimostratione come meno via meno faccia più)* which clearly echoes the title of Chap. XXII *On the contemplation of plus and minus, and that minus by minus makes minus* of *De regula aliza*. Although it is based on arguments very similar to Cardano's, Bombelli reached a diametrically opposite result, fully legitimizing the rule of signs.[34]

---

poterlo nominare: però si chiama Radice sorda, overo indiscreta, come sarebbe se si havesse a pigliare il lato di 20, il che non vuol dire altro, che trovare un numero, il quale moltiplicato in se stesso faccia 20; il ch'è impossibile trovare, per essere il 20 numero non quadrato" (Bombelli 1572, pp. 3–4).). The definitions of n-th root ($n = 3, 4, 5$) that follow are quite similar.

[31]There are only few authors presenting the geometric construction of the cube root of a given segment, e.g. Fibonacci, Pacioli and Tartaglia. On this topic see Rivolo and Simi (1998).

[32]On this aspect see, for example, Giusti (1992).

[33]"If in a right-angled triangle a perpendicular is drawn from the right angle to the base, then the straight line so drawn is a mean proportional between the segments of the base".

[34]As noticed by Ettore Bortolotti: "This chapter should rather be entitled: Proof of how it is necessary to put $- \cdot - = +$ so that the distributive property of the product remains valid" (Bortolotti and Forti 1966, p. 77, n.30).

Fig. 7.1 Construction of the
cubic side *fd* of a given
segment *de* (Bombelli
1572, Palat.8.5.2.1: p. 49).
Florence, National Central
Library

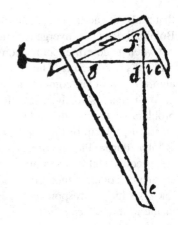

Treating binomials and cubic residuals, that is, expressions of the form $\sqrt[3]{a} \pm \sqrt[3]{b}$,
produces what is nowadays denoted as a cube root of a complex number, but what
Bombelli denoted as "another kind of linked cube root, which is very different from
the others and arises in the chapter dealing with the equation of form $x^3 = px + q$
when $p^3/27 > q^2/4$"[35]. Bombelli does not specify the nature of *linked cube roots*
$\sqrt[3]{a \pm \sqrt{-b}}$ ($b > 0$) but he does observe that radicals of the form $a \pm \sqrt{-b}$ do not
fulfill the usual rules of calculation, because the square root of a negative quantity
cannot be either negative or positive, being, as already glimpsed by Cardano, "some
recondite third sort of thing". This consideration forces Bombelli to invent new
signs, rather than new numbers, and to establish proper composition rules for them.
In the very famous passage that follows, the author highlights the urgent necessity of
being able to manipulate these linked cube roots, in order to overcome the obstacle
of the irreducible case, very frequent in the resolution of third and fourth degree
equations

> this kind of square root has in its calculation different operations than the others and has
> a different name. Since when $p^3/27 > q^2/4$, the square root of their difference can be
> called neither positive nor negative, therefore I will call it more than minus (più di meno)
> when it should be added and less than minus (meno di meno) when it should be subtracted.
> This operation is extremely necessary, even more than for the other *linked* cube roots which
> come up when we treat fourth-degree equations (complete or not) because the cases in
> which we obtain this [new] kind of root are many more than the cases in which we obtain
> the other kind. This new kind of root will seem to most people more sophistic than real; this
> was the opinion I held, too, until I found its plane geometrical proof [. . .] I will first treat
> multiplication, giving the law of plus and minus[36]

---

[35]"... un'altra sorte di Radici cubiche legate, molto differenti dall'altre, la qual nasce dal capitolo
di cubo eguale a tanti e numero, quando il cubato del terzo delli tanti è maggiore del quadrato della
metà del numero" (Bombelli 1572, p. 133). For an analysis of this remark and the irreducible case
in Bombelli, compare La Nave and Mazur (2002), Kenney (1989).

[36]The very common interpretation of this rule in terms of the imaginary unit i makes the
comprehension easier for the modern reader (and this is the reason why I write it in the square

Plus by more than minus makes more than minus	$[+1 \cdot +i = +i]$
Minus by more than minus makes less than minus	$[-1 \cdot +i = -i]$
Plus by less than minus makes less than minus	$[+1 \cdot -i = -i]$
Plus by more than minus makes more than minus	$[+1 \cdot +i = +i]$
Minus by more than minus makes less than minus	$[-1 \cdot +i = -i]$
Plus by less than minus makes less than minus	$[+1 \cdot -i = -i]$
Minus by less than minus makes more than minus	$[-1 \cdot -i = +i]$
More than minus by more than minus makes minus	$[+i \cdot +i = -1]$
More than minus by less than minus makes plus	$[+i \cdot -i = +1]$
Less than minus by more than minus makes plus	$[-i \cdot +i = +1]$
Less than minus by less than minus makes minus	$[-i \cdot -i = -1]$

However, in order to solve the irreducible case, a further step is needed, namely to "reduce" these particular "linked cube roots" to simpler expressions in order to manipulate them algebraically[37]; in other words, one must extract the cube root of $a \pm \sqrt{-b}$. The procedure developed by Bombelli is based on the assumption that, in applying the previous rule of signs, the cube of $x \pm \sqrt{-y}$ must remain an expression of the same form, that is

$$(x \pm \sqrt{-y})^3 = a \pm \sqrt{-b} \qquad (7.15)$$

where a and b are appropriate coefficients. If this equality is read in reverse, it suggests that the cube root of the expression $a \pm \sqrt{-b}$ is always of the form $x \pm \sqrt{-y}$, provided the conditions

$$\begin{cases} \sqrt[3]{a^2 + b^2} = x^2 + y^2 \\ a = x^3 - 3xy^2 \end{cases} \qquad (7.16)$$

---

brackets), but it is, as we will see, a forcing of Bombelli's way of thinking. It is also worth underlining that the symbol i was not introduced until the end of the eighteenth century. "La qual sorte di Radici quadrate ha nel suo Algorismo diversa operatione dall'altre e diverso nome; perché quando il cubato del terzo delli tanti è maggiore del quadrato della metà del numero, lo eccesso loro non si può chiamare né più né meno, però lo chiamarò più di meno quando egli si dovrà aggiongere, e quando si doverà cavare lo chiamerò men di meno, e questa operatione è necessarissijma più che l'altre Radici cubiche legate per rispetto delli capitoli di potenze di potenze, accompagnati con li cubi, o tanti, o con tutti due insieme, ché molto più sono li casi dell'agguagliare dove ne nasce questa sorte di Radici che quelli dove nasce l'altra, la quale parerà a molto più tosto sofistica che reale, e tale opinione ho tenuto anch'io, sin che ho trovato la sua dimostratione in linee [...] e prima trattarò del moltiplicare, ponendo la regola del più e del meno: Più via più di meno, fa più di meno; Meno via più di meno, fa meno di meno; Più via meno di meno, fa meno di meno; Meno via meno di meno, fa più di meno; Più di meno via più di meno, fa meno; Più di meno via men di meno, fa più; Meno di meno via più di meno, fa più; Meno di meno via men di meno, fa meno" (Bombelli 1572, p. 169).

[37]See the paragraph *Modo di trovare il lato cubico di simil qualità di radici* (Bombelli 1572, from p. 180 on). The same problem appeared even when the unknown was expressed as the sum or difference of the usual linked cube roots of the form $\sqrt[3]{a \pm \sqrt{b}}$. Both in the *Ars magna* and in the *Algebra* there are procedures aimed at rationalizing these expressions in particular cases.

are satisfied. In this system the unknowns x and y can be "gropingly" solved, as Bombelli himself states.[38]

At this point, the irreducible case is then completely solved, at least for the special cases where it is easy to find the unknowns x and y.

These preliminary remarks open the way for the discussion of cubic equations, which can be found in the second book of the *Algebra* and which is developed according to the 3-step scheme already followed by Cardano in his *Ars Magna*: a statement of the rule in rhetorical form; numerical examples and a geometric construction of the solution. We now focus our attention on the last of these aspects.

In the *Ars magna*, Cardano proposed geometric representations of the solutions of the three cases of cubic equations, based on re-reading, in terms of decomposition into cubes and parallelepipeds, Tartaglia's solution algorithm.[39] For example, in the case of the equation $x^3 + px = q$, if we assume that the quantities u and v ($u > v$) defined by the conditions

$$\begin{cases} u - v = q \\ uv = \frac{p^3}{27} \end{cases} \tag{7.17}$$

represent cubes, then the solution is given by the subtraction of their sides, that is $x = \sqrt[3]{u} - \sqrt[3]{v}$. Starting from the typical equation $x^3 + 6x = 20$, Cardano considers a cube of side $\sqrt[3]{u}$ and another, of side $\sqrt[3]{v}$ such that $u - v = 20$ and $uv = 8$ and he decomposes the larger into cubes and parallelepipeds in such a way that the segment $\sqrt[3]{u} - \sqrt[3]{v}$ satisfies the starting equation.

The two remaining cubic equations admit similar geometric representations, but in this case Cardano implicitly excludes the irreducible case, since the possibility of decomposition requires the discriminant $\frac{q^2}{4} - \frac{p^3}{27}$ to be positive. However, these are not the only geometric constructions proposed by Cardano.

In Chap. XII of *De regula aliza*, entitled *De modo demonstrandi geometrice aestimationem cubi et numeri aequalium quadratis*, the solution of the equation $x^3 + 192 = 12x^2$ is constructed, thus providing a particularly interesting example as the equation is easily transformed (by putting $x = y - 4$) into the irreducible cubic $y^3 = 48y + 64$. Inspired by the commentary of Eutocius to the second book of Archimedes' *Sphere and cylinder*,[40] Cardano shows that the solution of

---

[38]Applying this method, Bombelli obtains, for instance, $\sqrt[3]{2 \pm \sqrt{-121}} = 2 \pm \sqrt{-1}$, $\sqrt[3]{52 \pm \sqrt{-2209}} = 4 \pm \sqrt{-1}$. The first case deals with the cube roots obtained by the Cardano formula for the equation $x^2 = 15x + 4$: by summing $2 + \sqrt{-1}$ and $2 - \sqrt{-1}$ the *true* solution 4 is obtained. Compare (Bortolotti and Forti 1966, pp. 180–5).

[39]In the *Ars Magna*, Cardano attributes the solution formula to Scipione Del Ferro and Tartaglia, but strongly lays claim to the geometrical proof of the formula, which rigorously legitimizes the validity of the arithmetic algorithm.

[40]"Et hoc nos docet facere Eutocius Ascalonita in secundum de Sphaera et Cylindro bifariam, sed sufficiat adduxisse primam illius demonstrationem" (Cardanus 1570a, p. 25).

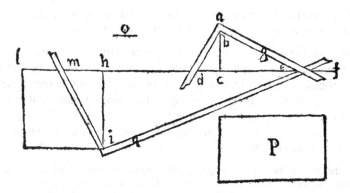

Fig. 7.2 Construction *in linea* of one cubic root (Bombelli 1572, Palat.8.5.2.1: p. 286). Florence, National Central Library

the equation can be represented as an intersection of a parabola and a hyperbola, but bitterly concludes that, although simple from the geometrical point of view, the construction is difficult to translate into arithmetical terms. Moreover, he adds, without any real justification, that he did not find the construction fully satisfactory; it can be conjectured that the impossibility of using only ruler and compass played a major role in this disappointment.[41]

When Bombelli, in his *Algebra*, faced the problem of geometric representation of cubic equations, he resumed the dialogue at a distance with the *Ars Magna* and *De regula aliza*.

The first of the two constructions for the case $x^3 + px = q$ essentially follows Cardano's method of decomposition into cubes and parallelepipeds and is even based on the same example, $x^3 + 6x = 20$. The second "in lines" ("in linea") construction, that is in two dimensions, instead recalls the "mechanical" method for the extraction of cube roots, since it again uses two sliding squares, arranged by trial and error, using a principle of continuity. For the equation $x^3 + 6x = 20$, the square *lhi* of side $hi = \sqrt{20}$ and the segment *hc* of length 6 perpendicular to one of its sides are constructed. After fixing the unit measure *dc*, the sliding squares are placed in such a way that one vertex coincides with the point *i* and the other slides on the segment *bc* perpendicular to *dc* so that the segments *bc* and *mh* remain equal. From simple Euclidean theorems Bombelli proves that *bc* and *mh* represent the solution of the given cubic (Fig. 7.2).

Unlike Cardano, Bombelli not only accepts a construction not executable by ruler and compass, but in this case considers it inevitable:

> and since it is known that there exists no real way to find two mean proportionals of two given lines, and it is necessary to fumble (as was shown in the case of the "in lines" extraction of cube roots), for this reason one should not account this proof worthless because

---

[41]"Et ideo facilis operatio Geometrica difficillima est arithmetice, nec etiam satisfacit" (Cardanus 1570a, p. 27). This construction is briefly shown in Maracchia (2003).

of the raising and lowering of the sliding square .g. so that .bc. becomes equal to .hm.. In fact when solid bodies are considered, it is not possible to do otherwise.[42]

The discussion of the case $x^3 = px + q$ begins with the statement of the rule followed by numerical examples; the irreducible case is introduced by an example where the special rule introduced by Cardano in his *Practica* ("la regola messa dal Cardano") is applied, consisting of adding to both sides a number that makes it possible to divide by a polynomial of the form $x \pm a$ to lower the degree of the equation.[43] If this manipulation is not possible, however, as in the equation $x^3 = 15x + 4$, the solution rule has to involve the new "linked cube roots":

> consider the third of "Tanti" [the coefficient of the linear term], that is 5, whose cube is 125, and subtract the square of the "number", which is 4, there remains -121. It will be denoted as "più di meno" and its square root $+$ of -11; its cubic side added to its residual gives $2 + $ of -1 and $2 - $ of -1, which put together give 4, and 4 is the required value. Even if this seems bizarre to many people, and this was also my opinion in the past, since it seemed to me more sophistic than real, nevertheless I have found the proof, which will be reported below, that can be provided in lines and the operations involved are of no difficulty, and many times the solution is found in numbers (as it has been found in this example).[44]

Bombelli observes that the geometric representation of a cubic equation through decomposition into cubes and parallelepipeds is possible only when it has a non-negative discriminant. When this does not happen, it is still possible to provide a geometrical proof of the existence of the root, as long as the use of sliding squares instead of ruler and compass is accepted, that is, the idea of determining a point in an approximate way has to be allowed.[45] Taking *ml* as a unit segment and *lf* of length

---

[42] ... e perché si sa che a trovare le due medie proportionali fra due linee date non ci è via reale, ma si opera a tentoni (come si è mostrato nella estrattione delle Radici cubiche in linea) però non si deve tenere questa dimostratione di poco valore per havere ad alzare et abbassare lo squadro .g. tanto che la .bc. sia pari alla .hm. perché dove intervengono corpi non si può fare altrimenti." (Bombelli 1572, pp. 287–288).

[43] The examined case is $x^3 = 12x + 9$; add 27 to both sides so that they are both divisible by $x + 3$.

[44] "Piglisi il terzo delli Tanti, ch'é 5, cubisi fa 125 e questo si cavi del quadrato della metà del numero, ch'è 4, resta -121. Il qual si chiamerà più di meno che di questo pigliata la Radice quadrata sarà $+$ di -11, che pigliatone il lato cubico ed aggionto col suo residuo fa $2 + $ di -1 et $2 - $ di -1, che gionti insieme fanno 4 e 4 è la valuta del Tanto. Et benché a molti parerà questa cosa stravagante, perché di questa opinione fui ancho già un tempo, parendomi più tosto fosse sofistica che vera, nondimeno tanto cercai che trovai la dimostratione, la quale sarà qui sotto notata, sì che questa ancora si può mostrare in linea, che pur nelle operationi serve senza difficultade alcuna, et assai volte si trova la valuta del Tanto per numero (come si è trovato in questo esempio). (Bombelli 1572, pp. 293–294).

[45] "this decomposition cannot be made in the mentioned way, but since it did not seem general to me, I investigated until I found a very general proof in a plane surface; since where bodies are considered mean proportional lines cannot be found if not by instruments; nobody should be surprised if this proof shows the same difficulty and if there was not such a difficulty, the invention of Plato and Archita from Taranto together with many other talented scientists about the duplication of the altar, that is a cube, would have been vain, (as widely treated by Barbaro in his Commentary on Vitruvius). Having the shield of so many talented men I will not strive to support the fact that the proof could not be carried out without involving the instrument

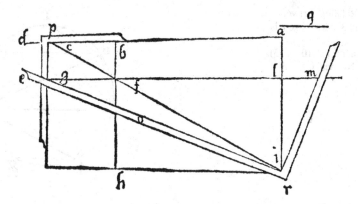

Fig. 7.3 Construction of one cubic root in the irreducible case (Bombelli 1572, Palat.8.5.2.1: p. 298). Florence, National Central Library

equal to the coefficient of the linear term 6 (the reference example is $x^3 = 6x + 4$), the rectangle *abfl* is constructed, whose area equals the constant term 4. The sliding squares are placed in such a way that the vertex of one is constrained to slide on the line *li* and to pass through *m* and the other in such a way that one arm can slide on the line *ad*; when the two arms intersect at the point *g*, we obtain a configuration where *li* represents the solution of the given cubic equation (Fig. 7.3).[46]

Ingeniously, Bombelli also suggested a way to geometrically represent expressions like $\sqrt[3]{a + \sqrt{-b}} + \sqrt[3]{a - \sqrt{-b}}$, going back to the cubic equation of which they are roots, and then proceeding to the "in lines" construction that we have just analysed.[47]

The legitimacy that Bombelli grants to this type of construction convinces him of the utility of Cardano's *radices sophisticae* and encourages him to determine appropriate rules of calculation, which however did not appear at all satisfactory to Cardano, who in the *Sermo de plus et minus* Cardanus (1663b)—the eloquently

---

[the sliding square]" ("tal agguagliatione non si potrà fare con detto taglio, però non parendo tale agguagliatione generale sono andato tanto investigando che ho trovato una dimostrazione in superficie piana generalissima, ma perché dove intervengono li corpi le linee medie non si possono ritrovare se non per via d'instromento, però non paia ad alcuno strano se questa dimostratione haverà la medesima difficultà, che quando non l'havesse saria stata vana la inventione di Platone ed Archita Tarentino con tanti altri valent'huomini nel voler duplare l'altare, overo Cubo (come largamente ne ha parlato il Barbaro nel Comento del suo Vitruvio), però havendo lo scudo di tanti valent'huomini non mi affaticarò in volere sostentar tal dimostratione non di potere far altramente che con l'instromento") (Bombelli 1572, p. 297).

[46]Given the triangle *mgi*, we deduce from the proportion $ml : li = li : lg$, with $ml = 1$ and $li = x$, that $lg = x^2$ and the area of the rectangle *rlg* is equal to x^3. Since the two rectangles *abfl* and *fgh* are equal, the rectangle *rlg* is also equal to $6x + 4$.

[47]In particular, the example offered by Bombelli is $\sqrt[3]{4 + \sqrt{-11}} + \sqrt[3]{4 - \sqrt{-11}}$ and the equation to which it belongs is $x^3 = 9x + 8$.

Fig. 7.4 Relations between
2-sides and 3-sides diagonals
(Cardanus
1570b, Magl.5.1.1110: p. 55).
Florence, National Central
Library

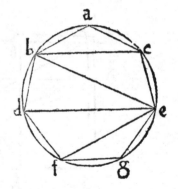

entitled work, written after reading Bombelli but published in 1663 by Spon—restated all his many perplexities with respect to Bombelli's results.

At the end of this analysis, we should recall that both Cardano and Bombelli emphasized the link between the irreducible case of cubic equations and the constructibility of certain regular polygons, the heptagon and nonagon, respectively (Fig. 7.4).

Proposition 66 of Cardano's *Opus novum de proportionibus* shows, in fact, that in a regular heptagon of side l, the 2-sides diagonal d_2 and the 3-sides diagonal d_3, satisfy the following continuous proportions

$$(l + d_3) : d_2 = d_2 : l \tag{7.18}$$

$$(d_2 + l) : d_3 = d_3 : d_2 \tag{7.19}$$

In other words, the side and the diagonals are in a special ratio named by Cardano *proportio reflexa*.[48] Then, taking l as unit segment and d_2 as the unknown side, the 2-sides diagonal may be expressed, with some algebraic transformation, by means of an irreducible cubic[49]

$$x^3 = \frac{7}{4}x + \frac{7}{8} \tag{7.20}$$

---

[48]"Propositio sexagesimasexta. Proportionem laterum eptagoni et subtensarum considerare et quae a reflexa proportione pendent" (Cardanus 1570b, pp. 55–56).

[49]The construction of the regular heptagon is one of the themes on which Ferrari and Tartaglia challenged each other in the *Cartelli* Masotti (1974): Tartaglia failed to propose a solution in the context of the challenge, but published the construction of the heptagon in (Tartaglia 1557, Part IV, Book I, c. 17). In Chap. XVI of the *De subtilitate*, especially since the edition of 1554, and in the *De proportionibus*, Cardano handles problems related to the construction of the regular heptagon. On this topic, see Field (1994) and Gavagna (2003).

In his *Commentaria in Euclidis Elementa*, a manuscript where various geometrical results were collected,[50] Cardano takes d_2 as unit segment and l as the unknown side, and comes to the irreducible cubic

$$x^3 + \frac{7}{27} = \frac{7}{3}x \qquad (7.21)$$

where $x = l + \frac{1}{3}$.

In question 135 of Book V of the *Algebra*, Bombelli deals with the problem of the construction of the regular nonagon inscribed in a circle of known diameter[51] and reduces it to the problem of trisecting an angle. Bombelli translates the geometrical problem into algebraic language and obtains the irreducible cubic $x^3 + 72 = 36x$ (where $2x$ represents the side of the nonagon). As previously mentioned, the versions of Books IV and V of the *Algebra* we now have, which were published only in 1929, date back to the 1550s, two decades before the final draft of the first three books published by the author in 1572. When he was writing Books IV and V, therefore, Bombelli had not yet untied the knot of the irreducible case and faced with the equation $x^3 + 72 = 36x$; he could only conclude disconsolately

> This problem seems impossible to me at present, and when the general way to solve the equation "cube and number equal to things" is eventually found it is difficult to imagine that in such a solution a cubic root will not appear. This suggests that, if a nonagon could be constructed, it could only be done by instruments, even though Oronce Finé and Albrecht Dürer have given rules for the construction of such nonagon, which are definitely false: and since the errors are clear, I will not explain them.[52]

7.3 From *Radices Sophisticae* to *Racines Imaginaires*

Even if Bombelli's *Algebra* could count on eminent admirers like Stevin and Leibniz, its spread within the scientific community was very limited and the *linked cube roots*, as well as the "new signs" invented by Bombelli, never underwent any

---

[50]The contents of the *Commentaria in Euclidis Elementa* (Par. Lat. 7217, Bibliothéque Nationale de France, Paris) and the meaning in Cardano's mathematical work are discussed in Gavagna (2003a).

[51]"It is the circle *abcdef*, whose diameter *be* is $\sqrt{192}$, inside which I would like to inscribe a regular nonagon; I was wondering about the length of one of his sides" ("Egli è il circulo *abcdef*, che 'l diametro *be* è $\sqrt{192}$ dentro del quale vorrei fare un nove faccie di lati eguali; addimandasi quanto sarà uno de detti lati" (Bortolotti and Forti 1966, pp. 639–641)).

[52]"Questa dimanda sino ad hora la tengo impossibile, fino a tanto che non sia retrovato il modo generale di agguagliare il capitolo di cubo et numero eguale a cose et dato che detto Capitolo ancor si ritrovi, dificil cosa sarà che in detto agguagliamento non intravenga qualche Radice cuba, che darebbe inditio, che potendosi formare detto nove faccie, non si potrebbe fare se non per via instrumentale, benché da Horontio et Alberto Duro siano state date regole da fare detto nove faccie, le quali sono falsissime: et per essere cose chiare non mi affaticarò in volerle dimostrare." Bombelli (Bortolotti and Forti 1966, pp. 639–641).

great development. In the early decades of the seventeenth century, however, the problem of computing the number of roots of an algebraic equation brought *radices sophisticae* back to the forefront.[53]

The question of the number of roots was not unrelated to Cardano or Bombelli, since the two gave it important consideration, although did not place it in a rigorous theoretical framework. In the first chapter of the *Ars Magna*, Cardano observed, even if his exposition is far from clear, that a third degree equation could have at most three roots, either *true* or *false*. In his *Algebra*, Bombelli noticed that for the irreducible cubic, the *linked cubic roots* always appeared in pairs[54] or, in modern terms, he realized that, at least in equations of third or fourth degree, the number of complex roots is even, since a root and its conjugate were always coupled together.

In the *Invention nouvelle de l'algebre* of 1629, Albert Girard studied the resolution of polynomial equations and stated, with great emphasis but without any proof, one of the first formulations of what is nowadays known as *Fundamental theorem of algebra*: "Any algebraic equation has as many solutions as is indicated by its maximum denomination [its degree], with the exception of incomplete equations".[55] The idea that guided Girard, perhaps borrowed from François Viète, consisted of the possibility of constructing a system of n equations in n unknowns for an equation of degree n, where the unknowns are the roots, and the equations express the relationships between the roots and the coefficients a_i (*factions*) of the equation $x^n + a_1 x^{n-1} + a_2 x^{n-2} + \cdots + a_n = 0$. Girard seems to suggest that the incomplete equation, in which one or more coefficients is equal to zero, may constitute an exception to his theorem because one or more relations could be missing; however, in his examples, he works correctly by annihilating the relations that correspond to the zero coefficient. The validity of his theorem is obviously subject to the legitimacy of expressions of the type $a \pm b\sqrt{-1}$—the only expressions that are taken into account—and, in fact, after showing that the solutions of $x^4 = 4x - 3$ are $1, 1, -1 + \sqrt{-2}, -1 - \sqrt{-2}$ Girard points out that

> To those who think that these solutions are impossible, I answer that they have to be accepted for three reasons: to ensure the general validity of the rule, because there are no other solutions, and for their utility.[56]

The third and last book of the *Géométrie* (1637) of Descartes, *On the construction of solid and supersolid problems* (*De la contruction des Problemes, qui sont*

---

[53]The historical development of this problem is described in Stedall (2011).

[54]"Notice that this kind of root cannot be obtained if not together with its conjugate" ("Si deve avertire che tal sorte di Radici legate non possono intravenire se non accompagnato il Binomio col suo Residuo").

[55]"Toutes les equations d'algebre reçoivent autant des solutions que la denomination de la plus haute quantité le demontre excepté les incomplettes" (Girard 1629, p. 45).

[56]The translation is not literal; the original passage is "Donc il se faut resouvenir d'observer tousjours cela: on pourroit dire à quoy sert ces solutions qui sont impossibles, je respond pour trois choses, pour la certitude de la reigle generale, & qu'il ny a point d'autre solutions, & pour son utilité" (Girard 1629, p. 47).

solides, ou plusque Solides), is devoted to this geometric constructibility. The terms *solid* and *supersolid* are borrowed from Pappus and denote those classes of problems that can be solved by conics and by "more complex" curves, respectively. Since the concept of simplicity of a curve is ambiguous,[57] Descartes proposes a classification of algebraic curves according to the degree of their equations and therefore he observes that "some general statements must be made concerning the nature of equations".[58] The starting assumption is a weak form of the *Fundamental theorem of algebra*

> Every equation *can have* as many distinct roots (values of the unknown quantity) as the number of dimensions of the unknown quantity in the equation (my italics).[59]

because Descartes, unlike Girard, did not consider negative solutions (*racines fausses*) of polynomial equations, and the degree of the equation becomes the upper limit of the number of roots.[60]

Descartes invents some *racines imaginaires* which are certainly not the *imaginary numbers* as thought of today, but rather, literally, some ghost-entities

> Neither the true nor the false roots are always real; sometimes they are imaginary, that is, while we can always conceive of as many roots for each equation as I have already assigned yet there is not always a definite quantity corresponding to each root so conceived of. Thus, while we may conceive of the equation $x^3 - 6x^2 + 13x - 10 = 0$ as having three roots, yet there is only one real root, 2, while the other two, however we may increase, diminish or multiply them in accordance with the rules just laid down, remain always imaginary.[61]

---

[57]"We should always choose with care the simplest curve that can be used in the solution of a problem, but it should be noted that the simplest means not merely the one most easily described, nor the one that leads to the easiest demonstration or contruction of the problem, but rather the one of the simplest class that can be used to determine the required quantity" ("Il faut avoir soin de choisir tousiours la plus simple, par laquelle il soit possible de le resoudre. Et mesme il est a remarquer, que par les plus simples on ne doit pas seulement entendre celles qui peuvent le plus aysement estre descrites, ny celles qui rendent la construction, ou la demonstration du Probleme proposé plus facile, mais principalement celles qui sont du plus simple genre, qui puisse servir a determiner la quantité qui est cherchée") (Smith and Latham 1954, pp. 152–155). See also Bos (1990) and Bos (2001).

[58]"il faut que ie die quelque chose en general de la Nature des equations" (Smith and Latham 1954, pp. 156–157).

[59]"Scachés donc qu'en chasque Equation, autant que la quantité inconnue a de dimensions, autant *peut il y avoir* de diverses racines, c'est a dire de valeurs de cete quantité". (Smith and Latham 1954, pp. 158–159).

[60]A similarly weak formulation was already contained in the *Arithmetica Philosophica* (1608) by Peter Roth, who instead excluded the imaginary roots from the sets of the acceptable ones and did not correctly compute the multiple ones. On the possible influence of the *Arithmetica Philosophica* on Descartes, compare Manders (2006).

[61]"Au reste tant le vrayes racines que les fausses ne sont pas toujours reelles, mais quelquefois imaginairea, c'est a dire qu'on peut bien tousiours en imaginer autant que iay dit en chasque Equation, mais qu'il n'y a quelquefois aucune quantité, qui corresponde a celles qu'on imagine, comme encore qu'on puisse imaginaire trois en celle cy $x^3 - 6xx + 13x - 10 = 0$, il n'y en a toutefois qu'une reelle, qui est 2, & pour les deux autres, quoy qu'on les augmente, ou diminue,

7.4 Conclusions

The solution formula of the cubic equation discovered around 1515 by Scipione Del Ferro and found a few years later by Tartaglia, confronted Cardano with the problem of studying expressions of the form $a \pm b\sqrt{-1}$, an obstacle that prevented him from giving a really general solution to equations of third and fourth degree. Before investigating foundational questions on the nature of *radices sophisticae* and their geometric representation, Cardano provided the rules for operating arithmetically with these strange expressions, but he immediately encountered the problem of establishing which sign they had. Although he realized that the quantities could not be considered negative or positive, but were "a third sort of thing", Cardano tried to give them a sign, even attempting to formulate a new rule of signs appropriate to his own needs. After noting the failure of this approach, Cardano tried to find a solution formula that did not contain roots of negative numbers, but his efforts, collected in *De regula aliza*, were not rewarded (and, evidently, could have never been rewarded).

In the attempt to place the new algebraic result into a more rigorous theoretical framework, Bombelli reconsidered the problem of the sign of expressions having the form $b\sqrt{-1}$ and introduced the signs "più di meno" and "meno di meno", for which he established appropriate rules of multiplication. On this basis, Bombelli founded an arithmetic of Cardano's sophistical quantities, allowing him to make sense of the irreducible case of cubic equations and, in the special cases where it was easy to extract the *linked cubic roots* $\sqrt[3]{a \pm b\sqrt{-1}}$, also allowing him to solve such equations, and obtain the real roots. According to Bombelli, this was not yet sufficient to determine the complete mathematical legitimacy of these quantities, which he was able to prove only when he was able to give a geometrical representation of the roots of an irreducible cubic, even if it was an approximated construction which fell outside the Euclidean spirit, to which Cardano had remained firmly connected. Bombelli's insights were not noticed or developed by the European mathematical community, and when sophistic quantities re-appeared on the scene, in the form of Girard's *solutions impossibles* or Descartes' *racines imaginaires*, they arose to ensure the validity of the so-called *Fundamental theorem of algebra*.

Thus, in the late Renaissance, both in the cases of Cardano and Bombelli as in those of Girard and Descartes, "complex numbers" were solutions to problems, that made sense of the solution formula for cubic equations and supported a basic result on the number of roots of a polynomial equation, respectively. However their formal properties were still unclear, and many later problems would be faced by mathematicians to see complex numbers as specific objects of study, giving them an

---

ou multiplie en la façon que ie viens d'expliquer, on ne sçauroit les rendre autres qu'imaginaires." (Smith and Latham 1954, pp. 174–175).

objective 'mathematical existence"[62] that would give them an undisputed place in nineteenth century mathematics.

Acknowledgements I would like to express my gratitude to Jackie Stedall, for competent remarks and generous help in translation, and to Nicole Jones, Arielle Saiber and John Stillwell for the careful reading of this essay. I would like to thank also Francesca Gallori of the National Central Library of Florence for her precious help and kindness.

References

Bombelli, R. (1572). *L'algebra. Opera di Rafael Bombelli da Bologna divisa in tre libri*. Bologna: Nella stamperia di Giovanni Rossi. http://mathematica.sns.it/opere/9/.Cited31Jul2013.

Bortolotti, E., & Forti, U. (Eds.), (1966). *L'algebra. Opera di Rafael Bombelli da Bologna*. Milano: Feltrinelli.

Bos, H. (1990). The structure of Descartes' *Géométrie*. In G. Belgioioso, G. Cimino, P. Costabel, & G. Papuli (Eds.), *Descartes: il metodo e i saggi* (pp. 349–369). Roma: Istituto dell'Enciclopedia Italiana (repr. in Lectures in the history of mathematics, American Mathematical Society, 1991).

Bos, H. (2001). *Redefining geometrical exactness: Descartes' transformation of the Early Modern concept of construction*. New York: Springer.

Cardanus, H. (1539). *Practica arithmetice sive mensurandi singularis*, Milan: Io. Antoninus Castillionaeus. http://bibdig.museogalileo.it/Teca/Viewer?an=000000966782.Cited31Jul2013.

Cardanus, H. (1545). *Artis magnae sive de regulis algebraicis liber unus* (2nd edn). Nüremberg: Johannes Petreius, 1570. http://bibdig.museogalileo.it/Teca/Viewer?an=000000300255. Cited31Jul2013.

Cardanus, H. (1570a). *De regula aliza libellus*. Basel: ex Officina Henricpetrina. http://archimedes.mpiwg-berlin.mpg.de/cgi-bin/toc/toc.cgi?dir=carda_propo_015_la_1570;step=thumb.Cited31Jul2013.

Cardanus, H. (1570b). *Opus novum de proportione numerorum, motuum, ponderum, sonorum aliarumque rerum mensurandarum*. Basel: ex Officina Henricpetrina. http://archimedes.mpiwg-berlin.mpg.de/cgi-bin/toc/toc.cgi?dir=carda_propo_015_la_1570;step=thumb.Cited31Jul2013.

Cardanus, H. (1663a). Ars magna arithmeticae. In *Hieronymi Cardani mediolanensis opera omnia* (Vol.4, pp. 303–376). Lion: Huguetan and Ravaud. http://www.cardano.unimi.it/testi/opera. html.Cited31Jul2013.

Cardanus, H. (1663b). Sermo de plus et minus. In *Hieronymi Cardani mediolanensis opera omnia* (Vol. 4, pp. 435–439). Lion: Huguetan and Ravaud. http://www.cardano.unimi.it/testi/opera. html.Cited31Jul2013.

Cardanus, H. (1968). *Ars magna or the rules of Algebra* (R. T. Witmer, Trans.). New York: Dover.

Confalonieri, S. (2013). The telling of the unattainable attempt to avoid the *casus irreducibilis* for cubic equations: Cardano's *De Regula Aliza*. With a compared transcription of 1570 and 1663 editions and a partial English translation. Ph.D. Thesis, Université Paris Diderot and Università degli Studi di Milano.

Cossali, P. (1996). La storia del caso irriducibile. In R. Gatto, (Ed.), *Transcription, introduction and notes*. Venezia: Istituto Veneto di Scienze, Lettere ed Arti.

Field, J. V. (1994) The relation between geometry and algebra: Cardano and Kepler on the regular heptagon. In E. Keßler (Ed.), *Girolamo Cardano: Philosoph, Naturforscher, Arzt* (pp. 219–242). Wiesbaden: Harassowitz Verlag.

---

[62]On the concept of "mathematical object" and its birth, see (Giusti 1999, pp. 87–93), where a chapter is entirely devoted to complex numbers.

Franci, R. (1985). Contributi alla risoluzione dell'equazione di III grado nel XIV secolo. In M. Folkerts, & U. Lindgren (Eds.), *Mathemata: Festschrift für Helmuth Gericke* (pp. 221–228). Stuttgart: F. Steiner Verlag.

Franci, R., & Pancanti, M. (Eds.), (1988). Anonimo (sec.XIV). Il trattato d'algibra dal manoscritto Fond. Prin. II.V.152 della Biblioteca Nazionale di Firenze, Quaderni del centro studi della matematica medievale n.18, Siena.

Gatto, R. (1992). Il caso irriducibile delle equazioni di terzo grado da Cardano a Galois. Atti e memorie dell'Accademia Nazionale di scienze, lettere e arti di Modena, Ser.VII, vol.X.

Gavagna, V. (2003). Alcuni aspetti della geometria di Girolamo Cardano. In E. Gallo, L. Giacardi, & O. Robutti. In *Conferenze e Seminari 2002–2003* (pp. 241–259). Torino: Associazione Subalpina Mathesis.

Gavagna, V. (2003a). Cardano legge Euclide: i *Commentaria in Euclidis Elementa*. In M. L. Baldi, & G. Canziani (Eds.) *Cardano e la tradizione dei saperi* (pp. 125–144). Milano: FrancoAngeli.

Gavagna, V. (2010). Medieval Heritage and New Perspectives in Cardano's *Practica arithmetice*. *Bollettino di Storia delle Scienze Matematiche, 30/1*, 61–80.

Gavagna, V. (2012). L' *Ars magna arithmeticae* nel *corpus* matematico di Cardano. In M. Massa Estéve, S. Rommevaux, & M. Spiesser (Eds.), *Pluralité ou unité de l'algébre à la Renaissance* (pp. 237–268). Paris: Éditions Honoré Champion.

Girard, A. (1629). *Invention nouvelle en l'algebre*. Amsterdam: Chez Guillaume Iansson Blaeuw.

Giusti, E. (1992). Algebra and geometry in Bombelli and Viète. *Bollettino di Storia delle Scienze Matematiche, 12/2*, 303–328.

Giusti, E. (1999). *Ipotesi sulla natura degli oggetti matematici*. Torino: Bollati Boringhieri.

Kenney, E. (1989). Cardano: "Arithmetic subtlety" and impossible solutions. *Philosophia mathematica, 2*, 195–216.

La Nave, F., & Mazur, B. (2002). Reading Bombelli. *The Mathematical Intelligencer, 24*, 12–21.

Malet, A. (2006). Renaissance notions of number and magnitude. *Historia Mathematica, 33*, 63–81.

Manders, K. (2006). Algebra in Roth, Faulhaber and Descartes. *Historia Mathematica, 33*, 184–209.

Maracchia, S. (2003). Algebra e geometria in Cardano. In: M. L. Baldi, & G. Canziani (Eds.), *Cardano e la tradizione dei saperi* (pp. 145–155). Milano: FrancoAngeli.

Masotti, A. (Ed.), (1974). *Ludovico Ferrari e Niccolò Tartaglia, Cartelli di sfida matematica*. Brescia. http://mathematica.sns.it/opere/24/.Cited31Jul2013.

Nenci, E. (Ed.), (2004). *Girolamo Cardano, de subtilitate*. Critical edition of Books I–VII. Milano: FrancoAngeli.

Pacioli, L. (1494). *Summa de arithmetica geometria proportioni et proportionalita*. Venezia: Paganino de Paganini. http://bibdig.museogalileo.it/Teca/Viewer?an=000000300039. Cited31Jul2013.

Rivolo, M. T., & Simi, A. (1998). Il calcolo delle radici quadrate e cubiche in Italia da Fibonacci a Bombelli. *Archive for History of Exact Sciences, 52*, 161–193.

Saiber, A. (2014). (monographic study in progress), *Well-versed mathematics in early modern Italy* (pp. 1450–1650).

Smith, D. E., & Latham, M. L. (Eds.), (1954). *The Geometry of René Descartes*. New York: Dover.

Stedall, J. (2011). *From Cardano's Great Art to Lagrange's reflections: filling a gap in the history of algebra*. Zürich: European Mathematical Society.

Tanner, R. C. H. (1980). The alien realm of the minus: deviatory mathematics in Cardano's writings. *Annals of Science, 37*, 159–178.

Tartaglia, N. (1546). Quesiti et inventioni diverse de Nicolo Tartalea Bresciano. Venturino Ruffinelli, Venezia.http://mathematica.sns.it/opere/27/.Cited31Jul2013.

Tartaglia, N. (1557–60). *General Trattato de' numeri et misure*. Venezia: Curzio Troiano. http://mathematica.sns.it/opere/22/.

Wagner, R. (2010). The nature of numbers in and around Bombelli's *L'algebra*. *Archive for History of Exact Sciences, 64*, 485–523.

Chapter 8
Random, Complex, and Quantum

Artur Ekert

I always found it an interesting coincidence that the two basic ingredients of modern quantum theory, namely probability and complex numbers, were discovered by the same person, an extraordinary man of many talents, a gambling scholar by the name of Girolamo Cardano. In his autobiography, *De vita propria liber* (Cardano 1643), written when he was 74, he described himself as "... hot tempered, single minded, and given to woman, ... cunning, crafty, sarcastic, diligent, impertinent, sad and treacherous, miserable, hateful, lascivious, obscene, lying, obsequious,.." and "... fond of the prattle of old men." In the chapter dedicated to "stature and appearance" we learn that he was a man of medium height with narrow chest, long neck, and exceedingly thin arms. His eyes were very small and half-closed, and his hair blond. He had high-pitched and piercing voice, and suffered from insomnia. He was afraid of heights and "... places where there is any report of a mad dog having been seen." The narrative veers from his conduct, appearance, diet, and sex life to meetings with supernatural beings and academic intrigues. A patchy but surprisingly readable account of a mind-set of a Renaissance man.

In order to see the fusion of his two discoveries, namely probabilities and complex numbers, let us briefly comment on each of them.

8.1 Probability

Cardano lived up to his reputation of a clever gambler. He knew that cheating at cards and dice was a risky endeavor, so he learned to win "honestly" by applying his discoveries concerning probabilities (Fig. 8.1). His *Liber de ludo aleæ* (The book on

A. Ekert (✉)
Mathematical Institute, University of Oxford, Oxford, UK

Centre for Quantum Technologies, National University of Singapore, Singapore
e-mail: artur.ekert@qubit.org

R. Lupacchini and A. Angelini (eds.), *The Art of Science*,
DOI 10.1007/978-3-319-02111-9_8,

Fig. 8.1 Caravaggio: *The Cardsharps, c.* 1594. Fort Worth (Texas), The Kimbell Art Museum. Two cheats and one dupe, beautifully painted by the young Caravaggio. One cheat, who concealed extra cards behind his back, plays with the unworldly boy while his accomplice peeps at the victim's hand and signals with his fingers. This was Cardano's world

games of chance) is a compilation of his scattered writings on the subject, some of them written as early as 1525, some of them later, around 1565 or so. The resulting treatise consists of 32 short chapters that were rescued from the pile of posthumous manuscripts, collated and included in the magnificent edition of Cardano's extant works, ten large folio volumes, published in 1663. It contains the first study of the principles of probability, the first attempt to quantify chance.

Of course, games of chance and the drawing of lots were discussed in a number of ancient texts and a number of mystics, loonies, and mathematicians enumerated the ways various games can come out. The snag was, most of these enumerations were not enumerations of equally likely cases, so they could hardly be used to calculate odds in a systematic way. Cardano was more careful. He started with the notion of fairness or, as he put it, "equal conditions":

> The most fundamental principle of all in gambling is simply equal conditions, e.g. of opponents, of bystanders, of money, of situation, of the dice box and of the die itself. To the extend to which you depart from that equity, if it is in your opponents favour, you are a fool, and if in your own, you are unjust.

In the simplest case of two players with equal stakes, the game is fair if the number of favorable and unfavorable outcomes is the same for each player. More generally, Cardano argued, fairness requires that the stakes in an equitable wager should be in proportion to the number of ways in which each player can win. He then went on to find fair odds for wagering with dice. He correctly enumerated the various possible throws, i.e., 6 for one die, 6×6 for two dice, and $6\times6\times6$ for three dice. For example, when discussing the case of rolling two symmetric dice he wrote

> ... there are six throws with like faces, and fifteen combinations with unlike faces, which when doubled gives thirty, so that there are thirty-six throws in all, ...

Trivial? Perhaps, but, for the time, Cardano showed remarkable understanding that the outcomes for two rolls should be taken to be the 36 ordered pairs rather than the 21 unordered pairs. In contrast, as late as the eighteenth century the famous French mathematician Jean le Rond d'Alembert (1754), author of several works on probability, made a silly mistake claiming that when a coin is tossed twice the number of heads that turn up would be 0, 1, or 2, which he viewed as three equiprobable outcomes. Cardano chose the correct sample space for his dice problems and effectively defined probability, or the odds, if you wish, as an appropriate ratio of favorable and unfavorable cases. For example,

> If therefore, someone should say, I want an ace, a deuce, or a trey, you know that there are 27 favourable throws, and since the circuit is 36, the rest of the throws in which these points will not turn up will be 9; the odds will therefore be 3 to 1.

Here the "circuit" is the number of possible elementary outcomes, that is, the size of the sample space, and the favorable outcomes are all throws which result in at least one face showing one, two, or three points. In other parts of the text he also quantifies odds as a ratio of favorable to all possible cases.

Cardano's careful enumerations provided, at the very least, good explanations why certain number of points were more advantageous than others. This was something many dice players had known from their experience, and even though they could relate it to the number of ways the throws can come out their counting was very problematic. For example, it was known, and regarded as puzzling,[1] that in a throw of three dice the sum of points is more likely to be 10 than 9, even though there are six ways in which the sum can be nine

$$1+2+6, \quad 1+3+5, \quad 1+4+4, \quad 2+2+5, \quad 2+3+4, \quad 3+3+3, \qquad (8.1)$$

and there are also six ways for the sum to be ten,

$$1+4+5, \quad 1+3+6, \quad 2+4+4, \quad 2+2+6, \quad 2+3+5, \quad 3+3+4. \qquad (8.2)$$

The fact that the outcomes should be taken to be ordered triples (27 of which sum up to ten but only 25 to nine) was not well understood. Thus even if Cardano's discussion had been limited to calculating the correct chances on dice, astragals, and cards, it could have been regarded as a great achievement, but he went further than that. He made several insightful general statements about the nature of probability. For example, he realized that when the probability of an event is p, then by a

---

[1]Galileo Galilei was explicitly asked, by one of the gambling noblemen at the court in Florence, to explain this puzzle, and so he did in his brief *Considerazioni sopra il Giuoco dei Dadi*, written around 1620.

large number n of repetitions the number of times the event will occur is not far from np. Although claims that he anticipated the laws of large numbers are difficult to justify, it is clear that his intuition was leading him in the right direction. The most remarkable part of *Liber de ludo aleæ* is Cardano's discussion of the probabilities for repeated throws of dice. It led him, after few unsuccessful attempts, to the correct power formula; given the probability p of a success in a single trial the probability of n successes in n independent trials is p^n. We can follow the process of his discovery in the text as it goes by trial and errors and he did not hide the errors; on the contrary, they are brought to reader's attention by chapter headings such as "On an Error Which I Made About This."

All this was written more than a century before a certain Chevalier de Méré, an expert gambler, consulted Blaise Pascal (1623–1662) on some "curious problems" in games of chance. Pascal wrote to his older colleague Pierre de Fermat (1601–1665), and it was through their correspondence, as we are often told, the rules of probability were derived. The thing is, *Liber de ludo aleæ* (1663) appeared in print over eighty years after Cardano's death and about nine years after Pascal's first letter. Thus, it is reasonable to assume that it had no impact on the subsequent development of the subject. However, in all fairness, one should recognize the fact that Cardano was the first to calculate probabilities correctly and the first to attempt to write down the laws of chance. According to Øystein Ore (1953), a Norwegian mathematician who elucidated many obscure parts of Cardano's gambling studies, it would be more just to date the beginning of probability theory from *Liber de ludo aleæ* rather than the correspondence between Pascal and Fermat. I certainly agree with that.

Cardano's "definition" of probability as a ratio of favorable to all possible outcomes is perfectly acceptable as long as you know (somehow) that all elementary outcomes are equiprobable. But how would you know? In many physical experiments the assumption of equiprobability can be supported by underlying symmetry or homogeneity. If we toss coins or roll dice we often assume they are symmetrical in shape and therefore unbiased. However, Cardano himself pointed out that "every die, even if it is acceptable, has its favoured side." No matter how close a real object resembles a perfect Platonic die, for mathematicians this approach is far from satisfactory for it is circular—the concept of probability depends on the concept of equiprobability.

You may be surprised to learn that the search for a widely acceptable definition of probability took nearly three centuries and was marked by much controversy.[2] In fact the meaning of randomness and probability is still debated today. Are there genuinely random, or stochastic, phenomena in nature or is randomness just a consequence of incomplete descriptions? What does it really mean to say that the probability of a particular event is, say, 0.75? Is this a relative frequency with which this event happens? Or is it the degree to which we should believe the event will happen or has happened? Is probability objective or subjective?

---

[2]For more details, see David (1998).

Most physicists would *probably* (and here I express my degree of belief) vote for objective probability. Indeed, physicists even *define* probability as a relative frequency in a long series of independent repetitions. But how long is long enough? Suppose you toss a coin 1,000 times and wish to calculate the relative frequency of heads. Is 1,000 enough for convergence to "probability" happen? The best you can say is that the relative frequency will be close to the probability of heads with at least such and such probability. Once again, a circular argument. Is there a way out of this vicious circle?

If you are prepared to forget about the meaning of probabilities and focus on the form rather than substance then the issue was resolved in the 1930s, when Andrey Nikolaevich Kolmogorov (1903–1987) put probability on an axiomatic basis in his monograph with the impressive German title *Grundbegriffe der Wahrscheinlichkeitsrechnung* (Foundations of Probability Theory). The Kolmogorov axioms are simple and intuitive. Once you identify all elementary outcomes, or events, you may then assign probabilities to them. Probability is a number between 0 and 1, and an event which is certain has probability 1. These are the first two axioms. There is one more. Probability of any event can be calculated using a deceptively simple rule—the additivity axiom:

> Whenever an event can occur in several mutually exclusive ways, the probability for the event is the sum of the probabilities for each way considered separately.

Obvious, isn't it? So obvious, in fact, that probability theory was accepted as a mathematical framework theory, a language that can be used to describe actual physical phenomena. Physics should be able to identify elementary events and assign numerical probabilities to them. Once this is done you may revert to mathematical formalism of probability theory. The Kolmogorov axioms will take care of the mathematical consistency and will guide you whenever there is a need to calculate probabilities of more complex events. This is a very sensible approach apart from the fact that it does not work! Today, we know that probability theory, as ubiquitous as it is, fails to describe many common quantum phenomena. The main culprit, as we shall see soon, is our innocuous and "obvious" additivity axiom. In order to fix the problem we need another mathematical tool, namely, complex numbers. They were discovered as a by-product of a fascinating search for an algebraic solution to the cubic equation, and this brings us back to Cardano.

8.2 Complex Numbers

Today, with the benefit of modern mathematical notation, we write the cubic equation as

$$ax^3 + bx^2 + cx + d = 0 \tag{8.3}$$

with a, b, c, and d being given real numbers. The Renaissance mathematicians knew that one can get rid off the square term and reduce this equation to the "depressed form,"

$$x^3 = px + q. \tag{8.4}$$

The trick involved replacing x by $x - b/3a$. Of course, not a single Renaissance mathematician would write these equations in this way. They were usually described in words, for example, expression

$$x^3 = 8x + 3. \tag{8.5}$$

would have been written by Cardano as

cubus æqualis 8. rebus p̓. 3., $\qquad\qquad$ (8.6)

where the Latin *rebus* ("things") refers to unknown quantities. In Italian texts the unknown "thing" was *cosa* and for a time the early algebraists were known as "cossists." I should also mention here that in Europe negative numbers were not considered seriously until the seventeenth century, so in Cardano's time different versions of the cubic equation must have been written down depending on the signs of the coefficients. Given that negative numbers were treated with a bit of suspicion, so taking roots of the suspicious numbers must have been almost heretical. After all solving equations meant solving specific mercantile or geometric problems. Thus "things" were measurable entities and whenever solving the quadratic equations, such as $x^2 + 1 = 0$, led to the square root of a negative number and it was assumed that the problem was meaningless with no solutions. Cubic equations were different. Some of them had perfectly respectable solutions, which could be easily guessed, and yet the square roots of negative numbers popped up halfway through, in the derivations of these solutions, and there was no way to avoid or to ignore them. This, to say the least, was puzzling.

The general solution to the depressed cubic reads

$$x = \sqrt[3]{\frac{q}{2} + \sqrt{\Delta}} + \sqrt[3]{\frac{q}{2} - \sqrt{\Delta}}, \tag{8.7}$$

where

$$\Delta = \left(\frac{q}{2}\right)^2 - \left(\frac{p}{3}\right)^3. \tag{8.8}$$

With their confusing notation and their reluctance to accept negative numbers the Renaissance mathematicians initially failed to grasp that this is indeed the general formula, which solves all cubics not just some specific cases. The most intriguing case, known as *casus irreducibilis*, occurs when $\Delta < 0$, for it involves square roots of negative numbers and always leads to three real solutions. Cardano, after learning the "general" solution from Tartaglia, tried to make sense out of *casus irreducibilis*. In 1539 he raised the matter with Tartaglia only to learn that he had not "mastered the true way of solving problems of this kind." Tartaglia, it seems, had little understanding of his own solution. Cardano, to be sure, did not elaborate on this case in *Ars Magna* (1545) but he did not avoid it either. Take, for example, equation

$$x^3 = 8x + 3, \tag{8.9}$$

which appears in Chap. 13 of *Ars Magna* with the comment: "Solving $x^3 = 8x + 3$ according to the preceding rule, I obtain 3." That must have baffled any careful reader who tried to work it out "according to the rule" since it is a clear *casus irreducibilis* with $\Delta = -1805/108$ and the solution which could only be expressed as

$$x = \sqrt[3]{\frac{3}{2} + \sqrt{-\frac{1805}{108}}} + \sqrt[3]{\frac{3}{2} - \sqrt{-\frac{1805}{108}}}. \tag{8.10}$$

And how do you get 3 out of that? Cardano does not say. This is surprising. After all Cardano is hardly afraid of square roots of negative numbers. On the contrary, few chapters later he constructs explicit examples to show how to deal with them. The examples must have been motivated by the *casus irreducibilis*. He discusses a problem of finding two numbers which sum to 10 and such that their product is 40. The solution is, of course, $5 \pm \sqrt{-15}$, or rather

$$5. \, \dot{p}. \, R_x. \, \dot{m}. \, 15 \text{ and } 5. \, \dot{m}. \, R_x. \, \dot{m}. \, 15.$$

Finding it difficult to make sense out of such "numbers" Cardano took a purely instrumental approach. He noticed that if you are prepared to ignore the question of what the square root of minus fifteen meant, and just pretend it worked like any other square root, then you could check that these mathematical entities actually fit the equation. He wrote: "Putting aside the mental tortures involved, multiply $5 + \sqrt{-15}$ by $5 - \sqrt{-15}$, making $25 - (-15)$ which is $+15$. Hence this product is 40." Richard Witmer, in his translation of *Ars Magna*, points out that the Latin phrase used by Cardano, namely *dimissis incruciationibus*, can also be translated "the cross-multiples having cancelled out." The sentence would then read "Multiply $5 + \sqrt{-15}$ by $5 - \sqrt{-15}$ and, the cross-multiples having cancelled out, the result is $25 - (-15)$, which is $+15$. Hence this product is 40." No "mental tortures" in this version. In another book, *Ars Magna Arithmeticæ*, Cardano remarks that $\sqrt{-9}$ is neither $+3$ nor -3 but some "obscure third sort of thing" (*quaedam tertia natura abscondita*). This is how complex numbers were announced to the world.

It took another 30 years or so, for one of those baffled readers of *Ars Magna*, Rafael Bombelli (1526–1572), the son of a Bolognese wool merchant, to provide a clear discussion of *casus irreducibilis* and to set up formal rules that allowed to perform consistent calculations with complex numbers. Still, it was Cardano, slightly ahead of his time, who wrote them down and, suppressing his uneasiness, performed simple operations on them. Unfortunately he never took them seriously and commented that they were as refined as they were useless!

8.3 Probability Amplitudes

Is there a connection between complex numbers and probabilities? Yes, there is. Amazingly enough they unite in the best physical theory we have today—a superb description of the inner working of the whole physical world—the quantum theory.

Quantum theory asserts that probabilities are less fundamental than probability amplitudes, which are complex numbers α such that $|\alpha|^2$ are interpreted as probabilities. The Kolmogorov axioms of probability theory seem to codify our intuition about probabilities quite well, however, we have now an overwhelming experimental evidence that by manipulating probabilities alone we cannot describe our physical world. The main culprit is the additivity axiom—nature simply does not conform to it. Example? Here is a classic double-slit experiment that illustrates this.

Imagine a source of particles, say electrons, which are fired in the direction of a screen in which there are two small holes. Beyond the screen is a wall with a detector placed on it. If the lower hole is closed the electrons can arrive at the detector only through the upper hole. Of course, not all electrons will reach the detector, many of them will end up somewhere else on the wall, but given a location of the detector there is a probability p_1 that an electron emitted by the source reaches the detector through the upper hole. If we close the upper hole then there is a probability p_2 that an electron emitted by the source reaches the detector through the lower hole. If both holes are open it makes perfect sense to assume that each electron reaching the detector must have travelled either through the upper or the lower hole. The two events are mutually exclusive and thus the total probability should be the sum $p = p_1 + p_2$. However, it is well established experimentally that this is not the case.

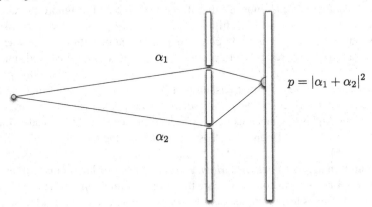

Quantum theory asserts that the probability of an event is given by the square of the modulus of a complex number α called the probability amplitude. Thus we associate amplitudes α_1 and α_2 with the two alternative events, namely "electron emitted by the source reaches the detector through the upper hole" and "electron emitted by the source reaches the detector through the lower hole," respectively. For consistency we must require that $|\alpha_1|^2 = p_1$ and $|\alpha_2|^2 = p_2$. However, and this makes quantum theory different, when an event can occur in several alternative ways, the amplitude for the event is the sum of the amplitudes for each way considered separately. In our case the amplitude that an electron reaches the detector when the two holes are open is

$$\alpha = \alpha_1 + \alpha_2 \tag{8.11}$$

and the associated probability

$$p = |\alpha|^2 = |\alpha_1 + \alpha_2|^2 = |\alpha_1|^2 + |\alpha_2|^2 + \alpha_1^\star \alpha_2 + \alpha_1 \alpha_2^\star \tag{8.12}$$

$$= p_1 + p_2 + |\alpha_1||\alpha_2|(e^{i(\theta_1 - \theta_2)} + e^{-i(\theta_1 - \theta_2)})$$

$$= p_1 + p_2 + 2\sqrt{p_1 p_2}\cos(\theta_1 - \theta_2). \tag{8.13}$$

where we have expressed the amplitudes in their polar form $\alpha_1 = |\alpha_1|e^{i\theta_1}$ and $\alpha_2 = |\alpha_2|e^{i\theta_2}$. The last term on the r.h.s. marks the departure from the classical theory of probability. The probability of any two mutually exclusive events is the sum of the probabilities of the individual events, $p_1 + p_2$, modified by what is called the interference term, $2\sqrt{p_1 p_2}\cos(\phi_1 - \phi_2)$. Depending on the relative phase $\phi_1 - \phi_2$, the interference term can be either negative (destructive interference) or positive (constructive interference), leading to either suppression or enhancement of the total probability p.

Note that the important quantity here is the relative phase $\phi_1 - \phi_2$ rather than the absolute values ϕ_1 and ϕ_2. This observation is not trivial at all. In simplistic terms— if an electron reacts only to the difference of the two phases, each pertaining to a separate path, then it must have, somehow, experienced the two paths. Thus we cannot say that the electron has travelled *either* through the upper *or* the lower hole, it has travelled through both. I know it sounds weird, but this is how it is.

Phases of probability amplitudes tend to be very fragile and may fluctuate rapidly due to spurious interactions with the environment. In this case, the interference term may average to zero and we recover the classical addition of probabilities. This phenomenon is known as *decoherence*. It is very conspicuous in physical systems made out of many interacting components and is chiefly responsible for our classical description of the world—without interference terms we may as well add probabilities instead of amplitudes.

Cardano had to go through the uncharted territory of complex numbers in order to obtain real solutions to cubic equations. As it happens, we do the same in quantum theory. We use complex amplitudes in order to calculate probabilities. The rules for combining amplitudes are deceptively simple. When two or more events are independent you multiply their respective probability amplitudes and when they are mutually exclusive you add them. This is just about everything you need to know if you want to do calculations and make predictions. The rest is just a set of convenient mathematical tools developed for the purpose of bookkeeping of amplitudes. But, as auditors often remind us, bookkeeping is important. Thus we tabulate amplitudes into state vectors and unitary matrices and place them in Hilbert spaces. We introduce tensor products, partial traces, density operators and completely positive maps, and often get lost in between.

8.4 Quantum Theory

You may ask whether we have to use Cardano's discoveries. Can we describe the world without probabilities and complex numbers? Well, let us try. Suppose you want to construct a framework theory, a meta-level description of the world, by armchair reasoning alone. Just pour yourself a glass of good wine, take a seat, take a sip, and think. How would you like to have your theory? It will be raw, for sure. And it should be as simple as possible. To start, assume that there exist physical systems which evolve from one state to another. What is this evolution? Should it be stochastic?

The concept of probability is useful no matter whether there are stochastic phenomena in nature or not. In the classical world, randomness arises as a consequence of incomplete description or knowledge of otherwise deterministic dynamics. Mind you, probability theory was developed by people who, by and large, believed that "things don't just happen." Cardano, with all his superstitions, was the borderline case but 200 years later Pierre-Simon Laplace (1749–1827) firmly believed that the world is ruled by causal determinism, i.e., every event is caused by, and hence determined by, previous events. Moreover, if at one time, we knew the positions and speeds of all the particles in the universe, then, at least in principle, we could calculate their behavior at any other time, in the past or future. This world view, known as predictive determinism, despite some practical difficulties, was basically the official dogma throughout the nineteenth century. It was challenged in the twentieth century by quantum theory which ruled out sharp predictions of measurement outcomes. The predictive determinism is unachievable, no matter how much we know and how much computational power we have we cannot make precise predictions of what is going to happen. Thus we are stuck with probabilities. Your armchair theory better be a statistical theory.

Classical probability is a good starting point—let us see where it leads. Assume that any physical systems can be prepared in some finite number of distinguishable states. Introduce the state vector which tabulates probabilities of the system being in a particular state and make sure that admissible transformations preserve the normalization of probabilities. Given any vector v with components $v_1, v_2, \ldots v_n$ the p-norm of v is defined as

$$(|v_1|^p + |v_2|^p + \ldots |v_n|^p)^{\frac{1}{p}} \tag{8.14}$$

thus for the probability vectors you make sure the 1-norm is preserved. Keep it simple, keep it linear, use transition matrices. Your admissible transformations are then represented by stochastic matrices P—they have nonnegative elements such that $\sum_m P_{mn} = 1$, i.e., entries in each column add up to one. The matrix element P_{mn} is the probability that the system initially in state labeled by n evolves over a prescribed period of time into the state labeled by m. The probability vector with components p_n evolves as $p_n \mapsto \sum_n P_{mn} p_n$.

Take a sequence of two *independent* evolutions, P followed by Q. What is the probability that the system initially in state n evolves over a prescribed period of time into the state m via some intermediate state k? The evolutions are independent so for any particular k the probability is $Q_{mk} P_{kn}$. But there are several intermediate perfectly distinguishable states k, thus there are several mutually exclusive ways to get from n to m. Following the Kolmogorov additivity axiom you add up the constituent probabilities, $\sum_k Q_{mk} P_{kn}$, and discover that the matrix multiplication QP in one swoop takes care of the multiplication and addition of probabilities. Products of stochastic matrices are stochastic matrices, so far so good.

Now you add one more requirement—*continuity* of evolution. Any product of stochastic matrices will give you a stochastic matrix but now you are asking for more. It should be possible to view any evolution as a sequence of *independent* evolutions over shorter periods of time. In particular, we should be able to take the square root, or the cube root, or any root of any transition matrix and obtain a valid transition matrix. Take, for example, a physical system with two states, say a physical bit, and consider a transformation which swaps the two states; a logical NOT if you wish, represented by the stochastic matrix,

$$\begin{pmatrix} 0 & 1 \\ 1 & 0 \end{pmatrix}. \tag{8.15}$$

Take the square root. The two eigenvalues of this matrix are ± 1 so you have to end with a matrix with complex entries, indeed

$$\begin{pmatrix} 0 & 1 \\ 1 & 0 \end{pmatrix} = \frac{1}{2}\begin{pmatrix} 1+i & 1-i \\ 1-i & 1+i \end{pmatrix} \frac{1}{2}\begin{pmatrix} 1+i & 1-i \\ 1-i & 1+i \end{pmatrix} \tag{8.16}$$

Square roots of stochastic matrices are, usually, not stochastic matrices. In other words—by adding the continuity requirement you gracefully thrashed your classical theory. Take another sip of wine and try again.

Keep state vectors and transition matrices T but let them have complex entries, simply because they pop up as soon as you start taking roots. Hopefully you will be able to relate complex numbers to probabilities later on. Take the continuity requirement seriously and parametrize transition matrices $T(t)$ with some real parameter t, that you may as well call time. Require that $T(t + s) = T(t)T(s)$, for any two time intervals t and s, and set $T(0) = \mathbf{1}$. Ha! This clearly points toward an exponential map $T(t) = \exp(tX)$, where X is any complex matrix. Now, taking the nth roots or inverses is a breeze: $T(t)^{1/n} = T(t/n)$ and $T(t)^{-1} = T(-t)$. You also recall that any matrix can be written in its polar form $T = RU$ where R is a positive matrix and U is unitary, it is analogous to writing a complex number in the polar form or viewing linear transformation T as the "stretching" R and the "rotation" U. But the exponential increase of stretching with time does not look good, you do not want to have exponential divergencies in your theory, so you had better drop R. Now, you are left with a unitary evolution of the form $U(t) = \exp itX$, where X is Hermitian. It looks good, decent periodic evolution, no exponential divergencies. But what does it mean? Now, you have to follow your hunch—probability should be preserved under the admissible evolution, so what is it that remains invariant under unitary operations. . . Eureka! The length of a vector! The Euclidian norm or the 2-norm, if you wish. Hence the squares of absolute values of complex components are probabilities. Now you have it all—state vectors with complex components, unitary transition matrices, and you know how to get probabilities out of the complex numbers. Congratulations, you guessed quantum theory without moving your butt from the armchair. Well, almost, there are a few holes in this plausibility argument, but they can be fixed, (with some more wine, of course).

More refined arguments can be found in a number of papers, in particular, in a very readable exposition by Lucien Hardy (2001), who argues, very convincingly, that if we try to construct a good statistical theory from a few (actually five) very reasonable axioms, then once we request *continuity* of admissible evolutions we will end up with quantum theory, and if this requirement is dropped we obtain classical probability theory.

The connection between amplitudes and probability is not trivial. Even the pioneer, Max Born (1926), did not get it quite right on his first approach. In the original paper proposing the probability interpretation of the state vector (wavefunction) he wrote:

> . . .If one translates this result into terms of particles only one interpretation is possible. $\Theta_{\eta,\tau,m}(\alpha, \beta, \gamma)$ [the wavefunction for the particular problem he is considering] gives the probability* for the electron arriving from the z direction to be thrown out into the direction designated by the angles α, β, γ . . ..
> * Addition in proof: More careful considerations show that the probability is proportional to the square of the quantity $\Theta_{\eta,\tau,m}(\alpha, \beta, \gamma)$.

Why do we square the amplitudes? Born's rule does not have to be postulated, it follows from the formalism of quantum theory. Here we usually refer to Gleason's theorem (1957). Although very helpful in clarifying the formalism and telling us what follows from what, the theorem itself offers very little in terms of physical insights and has no bearing on the issue of what probability is. There are more interesting and more productive approaches. For example, Scott Aaronson (2004) added a nice computer science flavor to the whole story by looking at a "what if" scenario. Suppose probabilities are given by the absolute values of amplitudes raised to power p. He showed that any linear operation that preserves the p-norm of a state vector is trivial apart from the two cases, namely, $p = 1$ and $p = 2$. For $p = 1$ we get stochastic matrices, that is classical stochastic evolution, and for $p = 2$ we get unitary matrices, that is quantum mechanics. In all other cases the only admissible operations are permutations of the basis vectors and sign changes and this may be not enough to account for our complex world.

As a realist—a true believer that science describes objective reality rather than our perceptions—I find Gleason's theorem too instrumental to my taste. After all, measurements are not just projectors but interactions between systems and measuring devices. Can unitary evolution alone shed some light on probability? Only then I would say that Born's rule really follows from the formalism of quantum theory. Fortunately, we do have a pretty good explanation of what probability really is in quantum physics. It was provided by David Deutsch (1999), with subsequent revisions by David Wallace (2003). They showed that no probabilistic axiom is required in quantum theory and that any decision maker who believes only in the non-probabilistic part of the theory, and is "rational" in the sense we already described, will make all decisions that depend on predicting the outcomes of measurements as if those outcomes were determined by stochastic processes, with probabilities given by Born's rule. All this follows from the bare unitary evolution supplemented by the non-probabilistic part of decision theory! One may argue about the status of decision theory in physics, however, by any account this is quite a remarkable result. It shows that it does make sense to talk about probabilities within the Everett interpretation and that they can be derived rather than postulated. And it is beautiful to see how a deterministic evolution of a state vector generates randomness at the level of an observer embedded and participating in the evolution.

Whatever the formalism, whatever the explanations, probability and complex numbers can hardly be avoided. It seems that we really need these Cardano's discoveries.

References

Aaronson, S. (2004). Is quantum mechanics an island in theoryspace? In: A. Khrennikov (Ed.), *Proceedings of the Växjö conference, Quantum Theory: Reconsideration of Foundations*, (quant-ph 0401062).

Born, M. (1926). Zur Quantenmechanik der Stoßvorgänge. *Zeitschrift für Physik, 37*, 863–867.

Cardano, G. (1545). *Artis magnae, sive de regulis algebrici liber unus.* In Cardano 1663. English tanslation by T. Richard Witmer: *The Rules of Algebra, (Ars Magna).* New York: Dover Books on Mathematics, 2007.

Cardano, G. (1643). *De propria vita.* Villery, Paris. (Jean Stoner, Trans.). *The Book of My Life (De vita propria liber).* New York: Review Books, 2002.

Cardano, G. (1663). *Opera Omnia Hieronymi Cardani, Mediolanesis*, 10 vols. Spoon, Lyons. All 10 volumes can be downloaded from the website of the Philosophy Department, University of Milan (http://www.filosofia.unimi.it/cardano/).

d'Alembert, J. (1754). Croix ou Pile. In *L'Encyclopédie*, vol. 4. Paris: Ed. Diderot.

David, F. N. (1998). *Games, gods and gambling. A history of probability and statistical ideas.* New York: Dover Publications.

Deutsch, D. (1999). Quantum theory of probability and decisions. *Proceedings of the Royal Society of London, A455*, 3129–3137. (quant-ph 9906015).

Gleason, A. (1957). Measures on closed subspaces of a Hilbert space. *Journal of Mathematics and Mechanics, 6*, 885–894.

Hardy, L. (2001). Quantum theory from five reasonable axioms. (quant-ph 0101012).

Ore, Ø. (1953). *Cardano, the gambling scholar*. Princeton: Princeton University Press. (Contains an English translation of *Liber de ludo aleæ* by Sydney H. Gould).

Wallace, D. (2003). Everettian rationality: defending Deutsch's approach to probability in the Everett interpretation. *Studies in the History and Philosophy of Modern Physics, 34*, 415–439. (quant-ph 0303050).

Index

R. Lupacchini and A. Angelini (eds.), *The Art of Science*,
DOI 10.1007/978-3-319-02111-9,
© Springer International Publishing Switzerland 2014

Printed in the United States
By Bookmasters